A GUIDE TO LABORATORY ANIMAL TECHNOLOGY

A Guide to Laboratory Animal Technology

Martin D Buckland NDP DipCG
Lynda Hall FIAT
Alan Mowlem FIAT MIBiol
Beryl F Whatley FIAT

William Heinemann Medical Books Ltd
London

William Heinemann Medical Books Limited
23 Bedford Square
London WC1B 3HH

First published 1981

ISBN 0 433 04590 6

Photoset in Great Britain by
D. P. Media Limited, Hitchin, Hertfordshire
Printed and bound in the United Kingdom by
Robert Hartnoll Ltd, Bodmin, Cornwall

Contents

Preface

A Guide to Laboratory Animal Technology has been written for animal technicians by animal technicians. The authors have between them amassed a range of experience with many species of laboratory animal in both research and breeding establishments. The aim is to provide an introduction and guide for the newcomer and student of the profession, and a useful comprehensive reference work for all those concerned with animals.

The common laboratory species are discussed individually with reference to anatomy, reproduction, maintenance, methods of handling and identification. For each species a list of common diseases is given. A comprehensive chapter on disease in general follows, which includes study of specific diseases previously listed. All aspects of laboratory animal husbandry have been considered including equipment, SPF techniques, transport and law in a way that can be readily understood. A guide to examination technique and a basic mathematics section are aimed to give help and confidence to the student with little experience in either field.

The authors have purposely omitted animal physiology as this specialised subject is well covered in other publications, many of which are listed in the appendixes under 'further reading'. However, among the many line drawings are diagrams of digestive and urinogenital systems of several species providing useful basic data and an easy reference for dissection.

Many photographs were specially commissioned, the authors taking particular care to illustrate methods of handling, sexing and various specialised techniques, a feature not usually found in a volume of this nature.

The appendixes contain much useful information, not only abbreviated facts from the relevant chapters but also comprehensive tables covering anatomical and physiological data for laboratory and farm animals.

Grateful thanks are extended to all those who, in many ways, have helped the authors to produce this book. We are indebted to Mrs Jacqueline Geens for the line diagrams and to Brian Hall ARPS for most of the photographs. Photographs and data from other sources are acknowledged in the text.

The animal unit and its equipment 1

The law governing the use and welfare of experimental animals has remained largely unchanged for over 100 years but is still appropriate today. By contrast it is only in the last 20 years or so that the accommodation in which laboratory animals are housed and caged has undergone significant change. In the early post-war years wooden rabbit hutches, heavy galvanised steel mouse cages and converted biscuit tins were commonly found in animal houses equipped with only the most elementary facilities and well away from the more prestigious laboratory buildings. Today, however, a more enlightened attitude prevails reflecting the importance of high quality animals to a modern research programme. The new generation of animal house and cage design has enabled animals to be bred and maintained free of many diseases that once were common-place; or indeed free of any microbiological association at all.

When planning a new unit or allocating space in an existing one, the needs of animals, animal technicians and research workers must all be taken into account. Animal units can be used exclusively for experimental or breeding purposes or for both and, according to their microbiological status, are termed conventional; barrier-maintained or minimal disease (MD); specified pathogen free (SPF), and germ free (GF or gnotobiotic). Specially designed facilities in which animals may be deliberately contaminated with microbes for experimental purposes are known as infected units. Radio-active materials may also be introduced, in which case the unit is known colloquially as a 'hot suite'. (SPF units are almost universally breeding units.)

Conventional units

In conventional units, apart from good basic hygiene, no special precautions are taken to prevent the entry of micro-organisms, and a change of clothing for staff and visitors is not generally obligatory. These units are still common in the UK and are usually of the older type with opening windows, unsophisticated heating and ventilation systems

and often originally served some other purpose. If such a unit houses experimental animals it is often not practicable to raise its status to minimal disease. The need for frequent access by the licensee must be acknowledged and it is pointless to insist on a strict entry routine which cannot be justified and may antagonise those using the facilities. It should be possible to achieve a standard of hygiene comparable to a well run commercial kitchen: rooms regularly cleaned and washed down with detergent, rinsed and disinfected; adequate hand-washing facilities using bactericidal soap and disposable towels; proper means of disposing of waste materials and facilities for equipment to be washed and where possible sterilised.

Minimal disease units

The current trend is to upgrade suitable existing units or build new ones to minimal disease status. These have air handling systems delivering filtered, temperature controlled air into the animal rooms at a rate of 15–20 changes per hour. The building has a higher internal air pressure than atmospheric pressure, termed positive pressure, to prevent airborne micro-organisms entering through cracks, for instance around exterior doors. Animal rooms are usually maintained at 12–25 mm (½ in to 1 in) water gauge pressure relative to the corridors which are in turn positive to the exterior by the same amount. Staff are required to change at least their outer clothing and footwear on entering. Pasteurised diet in stout plastic bags is supplied within a paper sack which is removed before the bags of diet are passed into the unit. Cages and bedding are autoclaved into the unit.

SPF units

The structure of the building is to a high standard to make it air-tight so that only sterile, filtered air is introduced to the animal rooms from the air conditioning plant. The rooms are at positive pressure to the corridors which are in turn positive to the atmosphere as in the minimal disease unit. All inanimate materials which enter the unit are sterilised, diet and other heat labile materials by irradiation, before being passed into the unit via a dunk tank containing a germicidal solution. Bedding, clothing and heat stable materials are autoclaved directly into the unit (see diagram of SPF unit autoclave on p. 20). Personnel enter via a shower and put on sterile clothing each time they enter. Waste materials are usually passed out through the autoclave and animals for issue through a special air-lock. In each case the chamber must be sterilised before the inner door is opened.

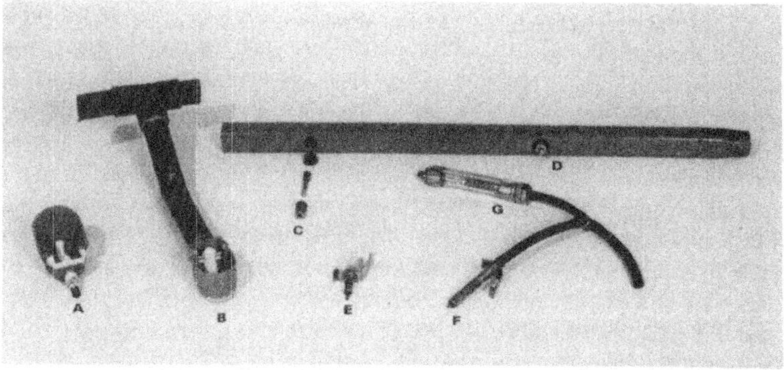

1.2. *Automatic watering equipment, (a) & (b) rear and front views of chicken drinker cup, (c) component parts of a chicken nipple drinker, (d) complete chicken nipple drinker* in situ *in supply pipe, (e) (f) small mammal nipple drinkers, (g) an in -line filter*

1.3. *Cabinet-type rack washer (AC Metal Fabrications Ltd)*

Caging

Cages vary from the small plastic 'shoe box' mouse cage to the large primate 'gang' or colony cage. Rodent cages are commonly made from polypropylene which is white or polycarbonate ('Makrolon') which is clear. Both are autoclavable but polycarbonate is affected by alkalis, some disinfectant solutions and autoclaving which all destroy the clarity of the plastic. Polycarbonate cages are about twice the price of polypropylene and while they will withstand severe impact they split easily when nested and stacked. The wire tops to these cages may be of stainless steel or galvanised mild steel. Stainless steel is a little more expensive than the galvanised but is lighter, much easier to clean, cannot rust and has a more attractive appearance. Cages may have solid bottoms or wire floors where faeces and urine fall into trays below. Only the tray requires frequent changing, the cage itself being changed less often, for example between litters or at the end of an experiment with a consequent saving in labour. Disadvantages of wire floors are that animals may trap their limbs and young animals may fall through.

Cages for housing animals such as rabbits, guinea-pigs, cats or primates are usually made of galvanised mild steel or anodised aluminium. The latter has the advantage of being lightweight, easy to clean and more resistant to corrosion. Other materials that may be encountered are polythene, glass-reinforced plastics (grp), wood or glass. (Figs 11.6 and 15.9)

Food may be supplied to rodents in a wire hopper in the lid of the cage when the large pellets are chewed between the wires. Most other species feed directly from a detachable hopper at the front of the cage. Water may be supplied from glass or plastics bottles fitted with various types of drinking nozzle or from a piped supply terminating in an automatic valve. Although the automatic system has the advantage that changing bottles two or three times a week is eliminated the nozzles have to be checked frequently to ensure that they are working, though in recent years they have become much more reliable. Automatic

1.4. *Tunnel-type cage washer (AC Metal Fabrications Ltd)*

systems could, perhaps, provide a route for cross infection. There are comprehensive arguments in favour of both systems but water bottles do indicate when animals are not drinking which may be an early warning of sickness. If it is necessary to give drugs in the drinking water then bottles must be used and it is as well to have a number in store for this purpose if an automatic system is normally employed. (Figs 1.1 and 1.2)

General considerations

The absence of windows in a room can markedly increase the wall space available for racking or cupboards. Windows in animal rooms, while of possible psychological benefit to staff, make environmental control very difficult: airflows are interrupted causing extremes of temperature throughout the year and making a uniform experimental environment impossible.

Insects may be excluded from the building by the use of filters on the air-conditioning intakes. The provision of an air-lock at the entrance incorporating an insecticide dispenser or electrical insect killer is recommended. Rodent barriers of sheet metal 45 cm (18 in) high with an outward facing top lip fitted across the entrance doors will exclude

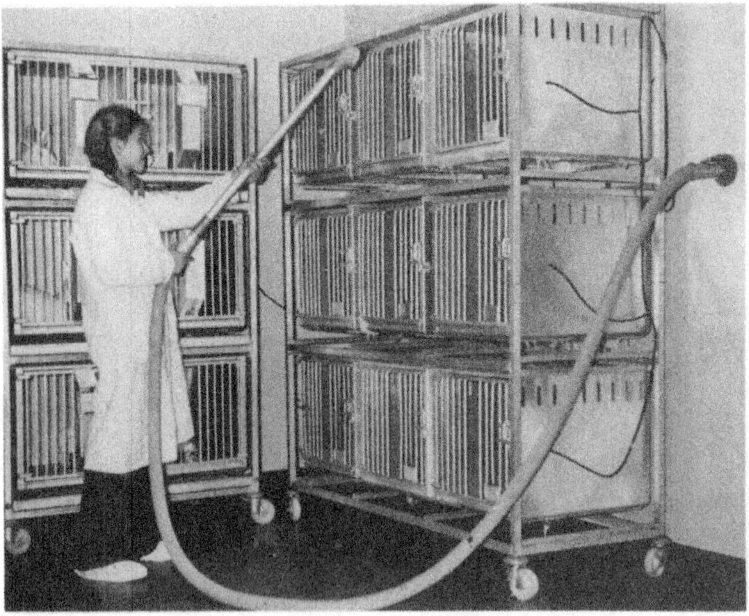

1.5. *Vacuum cleaning showing wall socket allowing machine to remain outside the animal room (National Institute for Research in Dairying)*

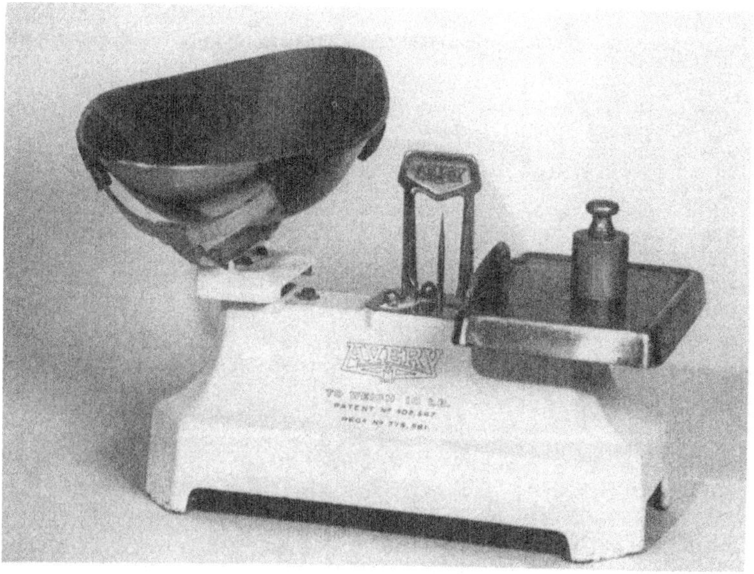

1.6. *Beam balance*

wild rodents and discourage non-essential visitors. Notices restricting entry and giving instructions for contacting unit personnel should be displayed at the entrance. A footbath, while of questionable use for removing micro-organisms, will deter the casual visitor.

Within any animal unit there should be adequate facilities for the storage of at least a week's supply of diet, clean cages, bedding and other stores. Facilities should be provided in an experimental unit where licensees may work undisturbed and without disturbing routine duties. This area should include storage for specialised experimental equipment.

Waste disposal and cleaning equipment

A considerable volume of waste material is produced by an animal unit. Dirty animal bedding forms the bulk and is usually bagged-up in the animal room or cage cleaning area and disposed of by incineration or collection by the Local Authority. Dead animals are usually incinerated or, where the volume of material is low, shredded in a specially designed machine and flushed into the sewerage system. This method should only be adopted after consultation with the local water authority who impose limits on the amounts of such material according to its biological oxygen demand (BOD) that can be discharged into the sewerage system.

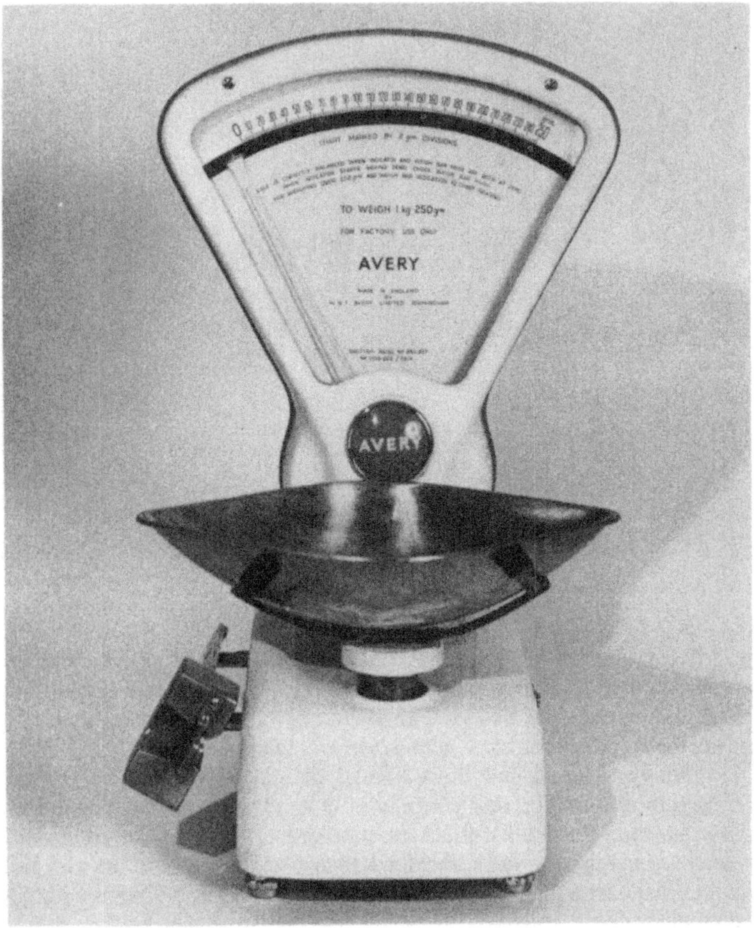

1.7. *Lever balance*

Small numbers of dirty cages are usually cleaned by hand, larger numbers by machine. These machines may be cabinet washers in which a batch of cages is washed at a time or a tunnel washer which operates as a continuous process. Each type has an automatic cycle which may include pre-wash, detergent wash and rinse stages. The number and size of cages and the frequency of washing will determine the appropriate choice (Figs 1.3 and 1.4). After being allowed to drain, cages may be autoclaved with or without bedding for re-issue to the animal rooms. The design and function of the autoclave is described on p. 18. Cages which are too large to pass through a cage washing machine may be cleaned by a jet of high pressure water, up to 8273 kN/m^2 (1200 psi), from a power washer, or may be steamed with a steam lance.

Cage washing equipment is expensive but is necessary as it is increasingly difficult to recruit suitable and sufficient labour to perform the work manually. In the long term with the rise in labour costs a machine may be more economical.

Room cleaning equipment

Dustpans, brushes and brooms should be provided for each room and should not be interchanged between rooms thus minimising the risk of transmitting disease. The equipment should be of plastic; wood and natural bristle are more likely to harbour micro-organisms. Plastic dustpans are quieter in use and do not dent when dropped and they usually have a sharper leading edge which enables a pile of dust to be picked up completely. The choice of mops is somewhat controversial; work carried out at the Central Public Health Laboratory at Colindale shows the traditional rag mop to be a positive breeding ground for bacteria unless autoclaved frequently whereas the cellulose sponge types do not last very long and inactivate some types of disinfectant. Floors may be cleaned by hosing down with water, the dirt being

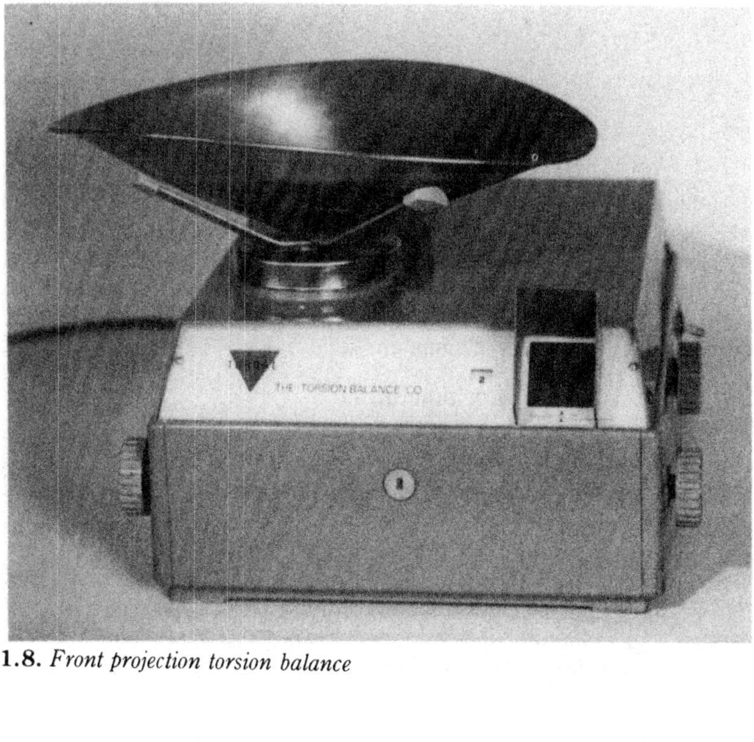

1.8. *Front projection torsion balance*

1.9. *Electronic balance*

loosened with a stiff broom and the water either swept into a drain with a rubber squeegee or removed with a wet-pickup vacuum cleaner. Both procedures leave the floor virtually dry reducing the hazard of slipping. In some modern animal houses a dry cleaning system using a vacuum cleaner may be preferred particularly in rabbit rooms where large quantities of loose fur accumulate. (Fig 1.5)

Balances and weighing machines

Animals and feedstuffs often need to be weighed and there is a variety of machines to do this. The simplest type of balance is the beam balance which has a place in the animal house for weighing dietary ingredients and has the advantage of being rugged enough to withstand sterilisation in an autoclave and can therefore be taken into pathogen-free and germ-free environments (Fig 1.6). A lever balance similar to those found in greengrocers' shops is robust, usually has an adjustable damper to moderate the effects of movements of the animal in the weighing pan and is easy to read. (Fig 1.7)

A more sophisticated mechanical balance uses torsion bars rather than levers and the indicated weight is usually optically projected onto a small ground glass screen at the front of the case. Unless the balance can be well damped the figures move rapidly across the screen as the animal moves about on the weighing pan making them impossible to read. The scale range may be low; 0–50 g is acceptable for weighing mice but of little use for rats as the tare control which allows objects heavier than 50 g to be weighed will need frequent adjustment (Fig 1.8).

1.10. *Spring balance with dial*

Electronic balances indicate the weight of an object by the change in resistance in an electric field when a magnet is pushed into that field by the loaded weighing pan. The weight is often indicated on an electronic digital display. The balance can be linked to a mini-computer which processes the weights and prints out onto paper or produces a punched code tape of totals or means. These weighing machines are relatively expensive but large numbers of animals can be weighed and recorded in a short time by one operator, and they are relatively robust, requiring little maintenance compared with other types of balance (Fig 1.9).

Spring balances of the fisherman's type are useful on occasions for

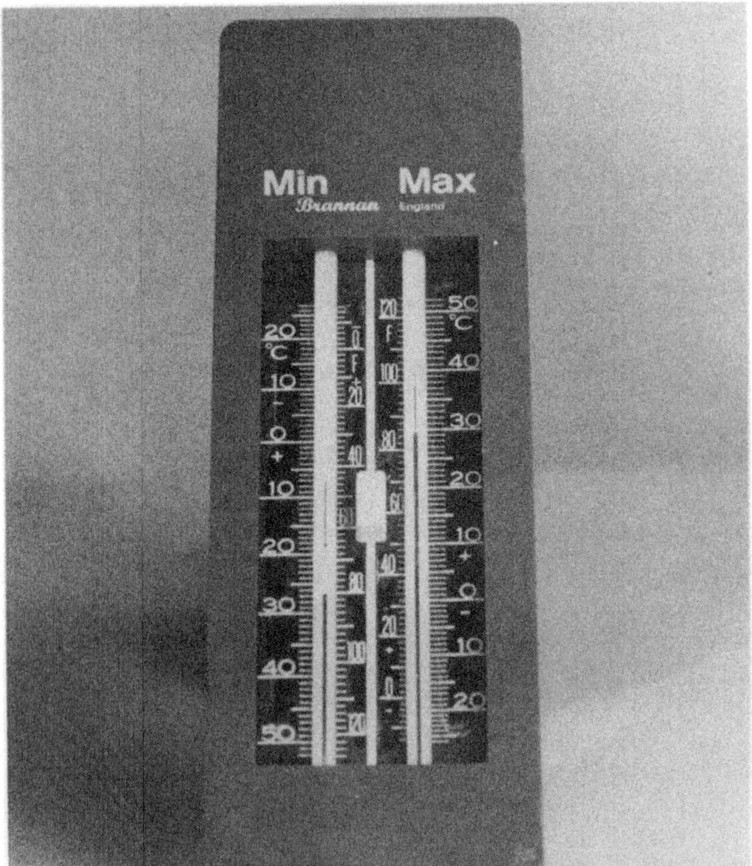

1.11. *Minimum -maximum thermometer*

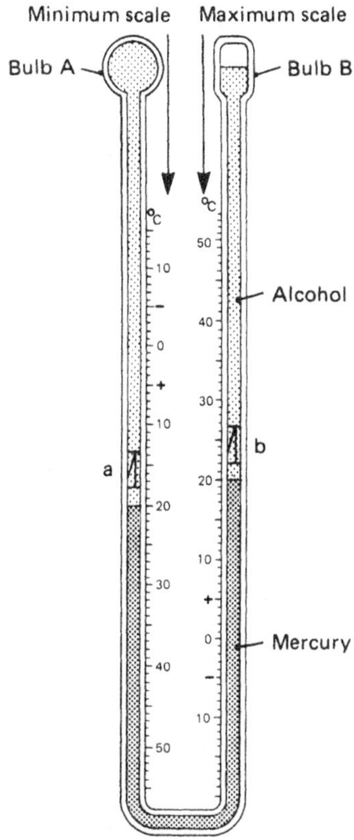

1.12. *Minimum and maximum thermometer. The minimum–maximum thermometer is an alcohol thermometer in which the expanding alcohol pushes against the column of mercury in the capillary tube. This in turn pushes the markers which indicate the lowest and highest room temperatures reached since the thermometer was last reset. As the temperature rises the alcohol in bulb 'A' expands and pushes the marker b up the tube. The bulb 'B' at the top of the tube allows space for the expansion of the alcohol. As the temperature falls the alcohol and mercury move back pushing marker a up and leaving the marker b in the tube where it is held by a weak spring. The ends of the markers nearest the mercury column indicate the highest and lowest temperatures reached. The ends of the mercury column indicate the present temperature and should read the same. The spring loaded markers are reset by drawing them back to the ends of the mercury column with an external magnet.*

weighing bulky or awkward objects. They may be fitted with a dial rather than the normal vertical scale. Adult chickens placed head downwards in a restraining cone are usually weighed on a spring balance (Fig 1.10).

When using a balance the following points should be observed.

1 Read the maker's instructions **before** using the balance and adhere to them.
2 Use the balance on a firm, vibration free surface and level the machine before use to avoid false readings. Check that the balance has not been disturbed since it was last levelled.
3 Check that the balance is zeroed and the tare adjustment correctly set. (Gross weight = article to be weighed plus container, nett weight = weight of article alone, tare weight = weight of container. Some balances may be adjusted to take account of the tare weight so that it is not necessary to subtract this from the gross weight for each weighing.)
4 When carrying the balance and after use make sure any locking device is in position and do not knock or jar the machine.
5 When not in use keep the balance covered with a plastic hood or bag.
6 Clean the balance after each period of usage.
7 Do not leave electric or electronic balances switched on for extended periods, for example, overnight.
8 Do not leave weights permanently on the scale or weight pans.

Temperature and humidity recorders

A standard thermometer will indicate the air temperature at the time

1.13. *Thermograph*

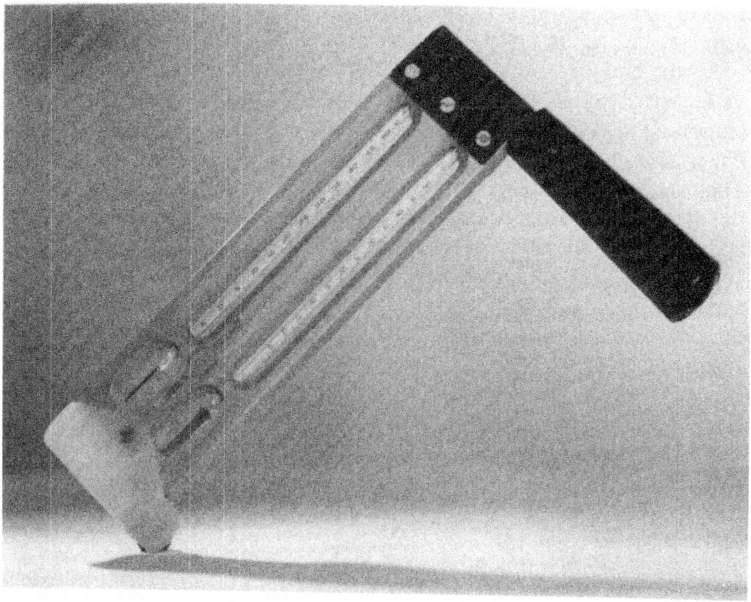

1.14. *Whirling hygrometer, cotton wick extending into water reservoir visible on lower thermometer*

of reading. A minimum–maximum thermometer will similarly show the actual temperature as well as the lowest and highest temperatures reached in any period since the indicators were last reset. (Figs 1.11 and 1.12)

A thermograph will provide a continuous record of temperature on a roll of graph paper, usually over a period of one week. The pen is moved up and down by the expansion and contraction of a bi-metallic coil. (Fig 1.13)

Relative humidity (RH) is an indication of the amount of water present in air and is expressed as a percentage of the ***maximum*** amount of water that the air is able to hold at *that particular temperature*. At 100% RH the air would be saturated: if any more water were added at that temperature mist would form as water droplets were precipitated. The higher the temperature of the air the more water it can hold. RH is the most commonly used measure of air humidity though absolute or specific humidity is sometimes used.

There are several instruments for measuring RH. A thermohygrograph looks very similar to a thermograph but has a second pen attached to strands of human hair which change length according to the humidity.

A hygrometer is a device where the RH is indicated on a dial by a pointer moved by a coil of special paper which reacts to the amount of water in the air by coiling or uncoiling.

A wet and dry bulb thermometer consists of two identical thermometers one of which has its bulb surrounded by a cotton wick immersed in a small pot of distilled water. As the water evaporates from the wick it takes heat from the bulb of the thermometer, so cooling it. The rate of evaporation is dependent on the amount of water present in the air. The temperatures indicated by both thermometers are noted and the RH determined by reference to standard tables. A more sophisticated instrument is the whirling hygrometer in which the wet and dry bulb thermometers are whirled through the air to obtain temperatures in moving air leading to a more accurate calculation of RH. (Fig 1.14)

Sterilisation and disinfection 2

Sterilisation is the complete removal or destruction of all living organisms including viruses, spores and vegetative forms. Sterilisation is an absolute term; nothing can be 'partially sterile' or 'nearly sterile' – it is either sterile or it is not. Following sterilisation an article must be protected from recontamination.

Note that disinfection is the removal or destruction of pathogenic (disease causing) bacteria excluding bacterial spores. Non-pathogenic bacteria may remain. In order that an article may be termed sterile an individual test must be carried out by the use of autoclave tape or spore strips, or by a sterilisation method which by design can guarantee sterility. Reference to these checks and tests will be made later.

Methods of sterilisation

There are two methods of sterilisation: (i) physical – which includes the application of heat in various ways, irradiation and filtration; (ii) chemical – usually by liquids or gases.

For any sterilisation process to be effective it is essential that surfaces to be treated are first thoroughly cleaned to remove all debris and grease which could protect organisms from the sterilising agent. The success of any sterilising process partly depends on the level of the microbial load present; there is far less chance of 1 organism in 100 surviving than 1 in 1000, and thorough cleaning beforehand will effectively reduce the number of organisms present. This procedure would not be adopted if it were suspected that dangerous pathogens were present: the techniques for dealing with infective material are outside the scope of this book and readers are referred to the texts given at the end of this chapter.

Physical methods

Heat

A very commonly used and effective agent which can be applied in either wet or dry forms.

Dry heat (i) Direct flame, for example in the heating of a wire inoculating loop as used in bacteriology to red heat in a Bunsen flame, or the use of a flame gun to heat the surface of metal cages. Flaming tends to buckle sheet metal and damage galvanising, and with large masses of metal the equipment is not heated sufficiently. However, this method may be useful in farm situations for large animal pens or other surfaces where other methods would be impracticable.

(ii) Hot air oven. Articles are placed in an oven heated by gas or electricity to 160°C (320°F), and this temperature is maintained for one hour before being allowed to fall to near room temperature. Hot air ovens are often used to sterilise glassware and the slow raising and lowering of temperature is necessary to prevent the glass cracking. Hot air sterilisation is useful for powders or materials such as cotton wool which would be damaged by moisture. Hot air ovens usually incorporate an internal fan to ensure uniform heat distribution throughout the cabinet. It should be noted that plastics which are capable of being autoclaved may warp or melt if sterilised in a hot air oven.

Wet heat (i) Boiling in water is often an effective method of sterilisation. Most micro-organisms are killed by boiling at 100°C (212°F) for 10 minutes but some species can withstand hours of exposure at this temperature. If sodium carbonate is added to the water to make a 2% solution the sterilising effect is considerably enhanced. Boiling is usually restricted to surgical instruments though sterile disposable syringes and instruments are now widely used.

(ii) Free steam, that is steam not under pressure, was widely used in the dairy industry for sterilising milk churns and may occasionally be used in animal houses. Items to be treated are placed in a large steel cabinet (preferably insulated), the door closed and steam introduced at the top and allowed to escape through a pipe at the bottom of the cabinet having first displaced the air inside. When the temperature of the exhaust steam reaches 100°C (212°F) the steam is allowed to flow for 1½ hours. In practice the temperature rarely exceeds 98°C (208°F) and the process therefore suffers from the same disadvantages as boiling, in that some organisms may survive.

(iii) Steam under pressure as used in an autoclave is a commonly used method of sterilisation in animal houses, so the process and equipment will be described in some detail.

When water is boiled in an open pan, the temperature of the water and steam produced is 100°C (212°F) as the water vapour pressure is equal to atmospheric pressure. If a tight fitting lid incorporating a safety valve is clamped onto a strengthened pan (as in a domestic pressure-cooker) the pressure created in the vessel will cause the water to reach a higher temperature before boiling and while the steam remains under pressure it too will maintain a temperature higher than 100°C (212°F). (The vapour pressure of water has risen and is equal to the pressure in the vessel.) The principle is exactly the same for an autoclave. The temperature of the steam rises in proportion to the pressure; 103 kN/m² (15 psi) produces a temperature of 121°C (250°F) and 207 kN/m² (30 psi) a temperature of 134°C (273°F). These temperatures are commonly used for sterilisation. (A method of calculating the approximate temperature in °C at a particular pressure is to add 106 to the pressure in psi, e.g. 15 psi + 106 = 121°C, 30 psi + 106 = 136°C (a reasonable approximation of the true figure 134°C).) The principle by which steam sterilises is as follows. Some of the heat that is applied to boil water is stored as latent (hidden) heat in the steam produced, though the temperature of the steam does not rise above the temperature of that water. This steam is on the phase boundary or borderline that separates water from steam. If there are no droplets of water present in the steam, it is said to be dry but saturated, in that it cannot hold any more water without droplets being formed, and must remain in this state to effect the sterilisation process. When the steam in the autoclave comes into contact with a cooler surface, for example an animal cage or clothing material, it condenses back into water and gives up its latent heat to that material. It is this heat which causes a rapid rise in the surface temperature and kills any organisms present. Water or air in the steam reduces this effect, similarly if the steam is very hot (superheated) it becomes dry heat and is much less efficient.

A simplified diagram of a double-ended autoclave is shown below and assuming the machine is installed in an SPF unit, the position of the wall is indicated. In conventional animal houses the positioning is often similar so that sterilised cages may be stored in a physically separate area to the cagewasher and dirty cages. In both situations the terms 'clean' and 'dirty' are often used. The reducing valve brings the high pressure steam in the main supply down to the required pressure in the autoclave for safety reasons and to prevent excessive temperatures which might damage items such as plastic cages. (The *Health and Safety at Work Act 1974* states that all pressure vessels must be inspected and tested for safety annually by an insurance company). The steam in the chamber is prevented from cooling by the insulating jacket of steam surrounding it. The jacket remains pressurised between cycles and therefore separate valves to the jacket and chamber are provided. The baffle plate over the steam inlet is to prevent an incoming jet of steam disturbing loose material such as sawdust. The perforated tray prevents absorbent materials becoming wet if there is water on the floor of the chamber when the cycle is completed. The thermometer is situated in the steam outlet and will only indicate 100°C (212°F) when all the air has been expelled from the chamber and pure steam is flowing. The

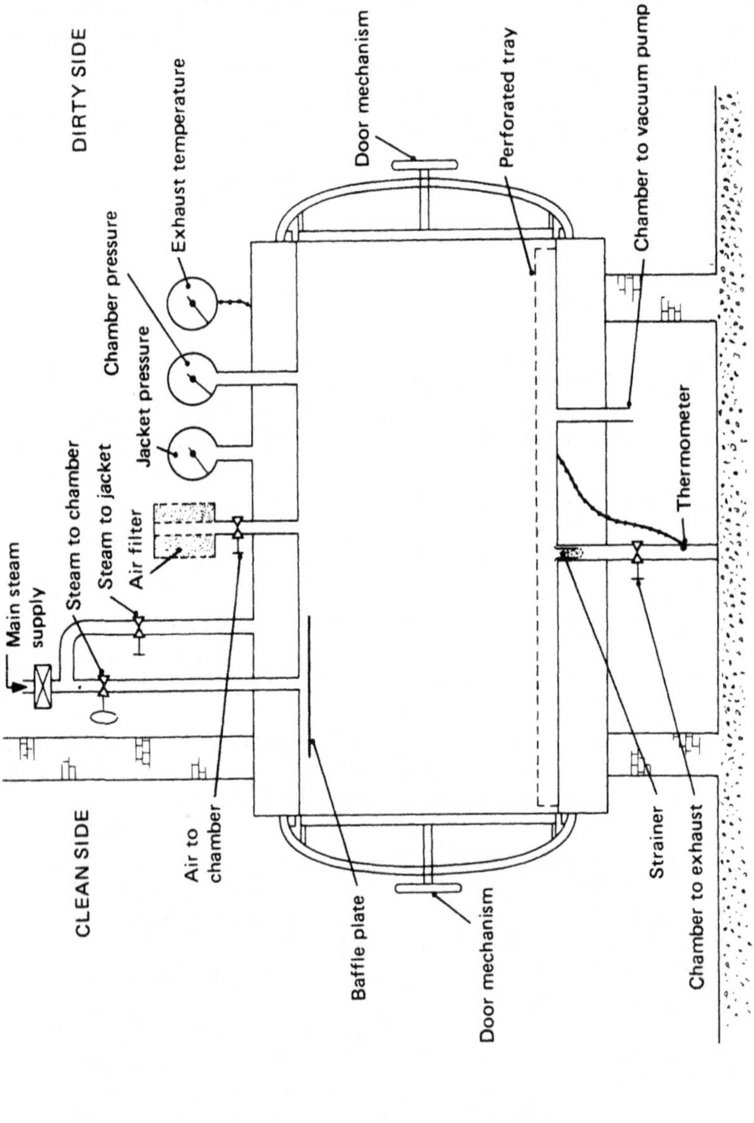

2.1. *Diagram of a double ended autoclave*

exhaust valve is closed and the temperature rises further as pressure increases until the required temperature is reached. The filter in the air inlet sterilises the air entering the chamber at the end of a vacuum cycle thus preventing re-contamination of the load. The door mechanism in modern machines is designed so that the cycle cannot start until the doors are securely locked and safe. A pressure of 103 kN/m² (15 psi) is capable of blowing an insecurely locked door weighing half a tonne off its hinges with possibly fatal results and it is therefore not wise to stand in front of an autoclave when the chamber is under pressure. No attempt must ever be made by anyone other than an authorised and qualified person to carry out any maintenance.

2.2. *Operating sequences of downward displacement and high vacuum autoclaves. The times given are approximate and for guidance only. Only a partial vacuum is pulled in some downward displacement machines, but these are gradually being replaced by automatic models.*

Downward displacement 103 kN/m² (15 psi)	Time taken (minutes)	**High vacuum** 103 kN/m² or 207 kN/m² (15 or 30 psi)		Time taken (minutes)
Operation		Operation		
Open 'steam to jacket' valve to pre-heat chamber at start of day	20	Same		20
Load chamber and close door		Same		
Open 'steam to chamber' valve, allow steam to displace air	20	Turn on vacuum pump to pull vacuum in chamber		4
When exhaust thermometer reads 100°C close 'exhaust' valve		Turn off vacuum pump when vacuum gauge reads 68–71 cm (27–28 in) Hg		
Allow chamber to fill with steam and wait till temp. and press. gauges read 121°C and 103 kN/m² (15 psi) respectively	15	Same except temp. may be 134°C and pressure 207 kN/m² (30 psi)		4
Start timing sterilising phase of cycle	20	Same except	121°C 134°C	15 3
Close 'steam to chamber' valve		Same		
Open 'exhaust' valve to reduce chamber pressure to 0 psi	2	Turn on vacuum pump and pull vacuum as before (If load to be dried add 7 mins)		4
Open door	4	Turn off vacuum pump		
		Open 'air to chamber' valve		2
		Open door		
Total	61		Total	17–29

There are two main types of autoclave in common use; the downward displacement and the high vacuum (Fig. 2.2). The downward displacement type relies on the fact that steam entering the top of the chamber will form a layer that will push the air out of the exhaust at the bottom (downward displacement of air by steam). The high vacuum machine has a steam ejector and/or pump to remove the air or steam from the load and the chamber creating a 98% vacuum which allows the steam to penetrate effectively to the centre of porous loads and reduces the time taken to reach sterilising temperature. At the end of the sterilising period a second vacuum is pulled to remove all the steam from the load so leaving it dry. This is important when bags of sawdust or clothing are being treated.

Some reasons for incomplete sterilisation.

1 Overloading – prevents free circulation and penetration of steam.
2 Incorrect loading – air trapped in solid bottom cages placed 'right way up'. They should be placed on their sides or upside down to allow the air to 'fall out', usually only a problem in downward displacement machines.
3 Use of wrapping materials that are impervious to steam.
4 Items not physically clean.
5 Faulty gauges (suspect if pressure and temperature gauges do not correspond).
6 Leaking valves or door seals (suspect if pressure and temperature gauges do not correspond).
7 Insufficient time at the sterilising temperature.
8 Contamination of load after sterilisation, for example, damaged or clogged air filter.

Methods of checking sterility. Either the conditions in the chamber are monitored during the cycle (live check) or the load is checked after the process has finished (post check). A check at the end of the cycle on an SPF unit autoclave has the disadvantage that non-sterile items could have been taken into the unit before it was apparent that the cycle was ineffective.

Live checks are made by using a thermocouple placed in the centre of the load and connected to a chart recorder on the outside of the machine, or by the readings of pressure, temperature and vacuum gauges. Post checks are carried out by placing one of the indicators described below in the load where penetration of the steam is most difficult, for example in the middle of a pile of clothing or a bag of sawdust. Browne's tubes are small glass tubes containing a heat-sensitive liquid which is red before being subjected to a given quantity of heat and appears green afterwards. A tube is placed in the centre of a load and, when removed at the end of the process, gives an immediate indication of whether the conditions for sterilisation were achieved. Spore strips or discs consist of absorbent paper impregnated with spores of *Bacillus stearothermophilus* which are killed only by an effective sterilisation process. A strip in a glycine envelope is placed in the centre of the load, removed at the end of the cycle and sent to a bacteriology laboratory where it is cultured for live organisms. If organisms are present the sterilisation process was ineffective. It is essential that an untreated strip from the same batch is also cultured as a control to ascertain that the organisms were viable.

Autoclave tape is a self-adhesive paper tape printed with diagonal stripes of heat-sensitive chemical which are barely visible prior to autoclaving but which appear brownish-grey after being subjected to a given temperature.

Radiation

Ultra-violet light (UV). This is not very often used in animal houses except for sterilising water supplies to SPF units. The water passes a battery of UV lamps, the light from which kills any organisms present. The flow rate of the water past the lamps must be slow enough to allow sterilisation to take place. The exit port in an SPF unit may also be sterilised by UV light after it has been used to pass animals or supplies to the outside. Some items may be passed into the unit by the same route but it should be remembered that any part of the article in shadow from the UV light will not be sterilised and items must therefore be turned to expose all surfaces to the light. The process is slow – two hours exposure minimum.

Ionising radiation. Two main types are used; high energy electrons known as cathode rays produced by a machine and, electromagnetic gamma rays, similar to short wave X-rays, produced by radioactive material such as ^{60}cobalt. In neither process do the materials sterilised become radioactive themselves. Cobalt is commonly used to sterilise supplies for SPF or germ-free units. While a dose of 2.5 megarads (Mrads) will sterilise diets for SPF units a second dose to give a total of 5.0 Mrads is usually given for supplies to germ free units to provide an extra margin of safety as the consequences of contamination in a germ-free isolator may be considerable both scientifically and financially.

The use of ionising radiation has the advantage that there is virtually no rise in temperature of the material sterilised which is particularly important when foodstuffs are treated. Disadvantages are that glass may darken, fabrics and plastics may be altered physically and the activity of drugs may be impaired.

Such irradiation is carried out by commercial contractors as the plant required is very expensive. Irradiated diets are available from millers, removing the need for the user to do his own packaging and transportation to and from the contractor. Small yellow indicator discs are stuck on the packages before passing through the irradiation plant and should on emerging have changed to dark red.

Filtration

In the laboratory fluids may be sterilised by being passed through special filters made of porcelain, asbestos, sintered glass or cellulose. A small self-contained filter that fits between the hub of a hypodermic needle and a syringe may be used to sterilise the fluid as it is injected into the body. Various types of air filter are used in the animal house. The air intake of the air conditioning system is protected by a roughing filter made of coarse fibres which excludes leaves and other gross debris from the system. The air is passed through a finer filter before being discharged into the animal room. The final filters on a system delivering air to an SPF unit are known as absolute or HEPA (High Efficiency Particulate Air) filters. Fine grade glass fibre mats rolled around a wire mesh cylinder are used on germ-free isolators and the air inlets of autoclaves.

Chemical methods. The definition of sterilisation needs to be borne in mind when using chemicals as many commercial preparations are only suitable for use as disinfectants. The following points should be considered when selecting a chemical sterilising agent.

1 Spores are only destroyed by chlorine, iodine, formaldehyde, glutaraldehyde and ethylene oxide (ETOX, ICI Chemicals Division, Macclesfield, Cheshire).
2 *Mycobacterium tuberculosis* is killed by several hours immersion in phenol.
3 Most viruses are killed by chlorine compounds, glutaraldehyde or formaldehyde.
4 Peracetic acid can be used in germ-free units to sterilise the entry port and the outside of packages to be passed into the isolator.
5 Quaternary ammonium or mercury compounds merely inhibit the reproduction of bacteria (bacteriostasis) and are therefore not sterilants.
6 In general chemicals take many hours to achieve their sterilising effect and some, including iodine, chlorine and the ampholytics, are inhibited by the presence of organic matter such as faeces, blood, grease, cotton, cork and some detergents.
7 Mercury compounds are neutralised by blood.

8 Chlorine and its compounds decompose during storage and corrode metals and fabrics.
9 Iodine stains and is expensive.
10 Peracetic acid and formaldehyde have extremely irritant vapours. (Formalin or formaldehyde BP is a 37% solution of formaldehyde gas dissolved in water and stabilised with methyl alcohol.)
11 Peracetic acid and formalin combine to form a carcinogen.
12 Glutaraldehyde and ethylene oxide have highly toxic vapours and therefore must be used with great care and in specialised equipment. (Modern autoclaves may be adapted to perform an ETOX sterilising cycle.) Ethylene oxide in air forms an explosive mixture and is therefore usually mixed with CO_2.

A general procedure for fumigating a room with formaldehyde is as follows:

Cages and other equipment that can be sterilised in an autoclave should be removed from the room (but if the room has been accidentally or deliberately contaminated as much equipment as possible should be left in the room to be fumigated). Walls, ceiling and floor should be thoroughly cleaned with detergent and hot water and rinsed with copious quantities of cold water. Care must be taken to prevent water entering electrical fittings such as lamps, switches, sockets, thermostats, telephones and other apparatus. Windows and ventilators should be sealed with tape – preferably from the outside of the room as areas under the tape will not come into contact with the gas. (If a loop of stout cord is placed under the taped-on covering of the inlet and exhaust grills of an air-conditioning system and the two ends left just inside the door the covers may be quickly ripped off the grills without the need to enter the vapour filled room at the end of the fumigation period).

The volume of the room should be calculated and the appropriate amount of vapour released into the room. The formaldehyde gas may be produced by the exothermic heating generated when two parts formalin are added to one part crystals of potassium permanganate (not potassium to formalin or a violent reaction will result). Formaldehyde may be produced by heating formalin in a pot fitted with a thermal cut-out device similar to that in an electric kettle or by heating prills (small beads) of a commercial preparation in a specialised heating device.

When liquid formalin is used 1 ml of a 10% solution for every cubic foot ($0.0283m^3$) of room space is required and when using prills the manufacturer's recommendations should be followed. Room temperature should be at least 18°C (64°F) and the relative humidity (RH) 80% but surfaces should not be running with water; the correct RH is the hardest condition to obtain but a bucket or two of hot water poured over the floor and allowed to evaporate is crude but reasonably effective.

Before the sterilisation process is started the maintenance and security departments should be informed of the hazard so that during the ventilation period after fumigation the air conditioning will not be turned off by any automatic thermostat or human intervention. Lengths of tape for sealing the door should be cut beforehand and prominent notices warning of the danger must be displayed.

Just before the vapour has been set to release the room should be vacated, the door closed and locked and the gaps around the frame and the keyhole should be sealed with the adhesive tape. Doors, windows and the points where pipes run through the wall to adjacent rooms should be checked for leaks. Sterilisation will probably be effected in a few hours but it is customary to leave the room sealed overnight or for a weekend. At the end of this period the room must be purged of formaldehyde gas. If the 'rip cord' method is used to pull the covers from the air-conditioning grills, the door to the room need only be opened momentarily but great care should be taken to avoid inhaling any fumes, which are also highly irritant to the eyes. The door should then be closed, locked, re-sealed and the warning notices left in place. As the inlet duct is still sealed air will be drawn in through gaps around the unsealed door thus preventing any escape of fumes from the room. If the room can only be ventilated by opening windows from the inside, breathing apparatus and suitable protective clothing must be worn by the person entering the room who should also wear a safety line held by two assistants outside the room so that rescue may be effected from the room in case of accident. The room should be left to ventilate for two or three days or until there is no trace of residual gas.

Methods of disinfection

One usually regards the application of liquid chemicals as disinfection but there are two other methods which are generally thought to be superior.

The first is the mechanical removal of organisms (cleaning) which, if done thoroughly, can remove 80–90% of contamination. The traditional scrubbing and washing of equipment and rooms is an example and should in any case precede disinfection. The second method of disinfection is the application of wet heat in one of the following ways:

(a) Autoclave (low temperature cycle)	80°C (176°F) for 10 min
(b) Free steaming	100°C (212°F) for 5 min
(c) Boiling	100°C (212°F) for 5 min
(d) Hot water immersion	65+°C (150+°F) for 10 min

The third method, application of chemicals, is probably the most used, misused and abused method. There are a large number of proprietary disinfectants available (see table on p. 192) many of which have different instructions for their use which are either separate from the container or go unread. Very often the claims made by the manufacturers are based on laboratory tests rather than the preferred 'in-use' test in which samples are tested in more realistic conditions.[1]

Chemical disinfectants may be grouped into broad classes such as phenols, alcohols, halogens, quaternary ammonium compounds ('quats') and the ampholytic agents. When choosing a chemical disinfectant the reader is advised to consult specialist texts. Sykes[2] reviewed the products available and Maurer[3] should also be consulted. Some of the factors affecting the efficiency of chemical disinfectants are listed below.

1 Anti-microbial range – may be narrow.
2 Time of exposure – may be critical.
3 Concentration – too high may cause skin irritation and is expensive, and lysol for example decreases in disinfective ability as the concentration rises above 5%.
4 Temperature. As this increases so does the efficiency of the disinfectant.
5 Inactivation by hard water, organic materials (faeces, blood and food), detergents, many plastics, cork, wood, rubber, cotton and nylon.
6 The ability of disinfectant to reach the organisms; detergent in the solution helps but may not be compatible with the disinfectant.
7 Numbers of organisms present.
8 Freshness of solution.
9 Many disinfectants lose activity on dilution.
10 Resistant organisms. Some microbes always survive and these give rise to resistant strains. Containers must be cleaned and sterilised between use otherwise resistant strains are favoured and become widespread.

Work done at the Public Health Reference Laboratory at Colindale has shown that the use of disinfectants with the traditional rag mop and bucket for cleaning floors may dramatically increase the bacterial count and spread infection to every point the mop touches. Hosing down or dry cleaning is much more effective.

Further Reading UFAW Handbook
IAT Manual

Reference [1]Kelsey and Maurer 1966. Bulletin of Ministry of Health and the Public Health Service 25.180 ('in-use' test)
[2]Sykes G (1967) *Disinfection and Sterilisation*, 2nd ed. Spon., London
[3]Maurer (1977) *Hospital Hygiene*, Edward Arnold

Transport of laboratory animals

3

The essence of transporting animals is to choose the quickest, safest and most efficient means of travel for the protection of the consignment. Any animal whether travelling 5 miles or 2000 must begin the journey in a completely healthy state. The fact that animals will be distressed by just being put into a box, gives an indication of the stresses they will have to endure on the journey.

A number of orders and amendments cover the transportation of animals on land, sea and air. The *Transit of Animals (General) Order 1973* provides general measures intended to safeguard the welfare of animals during their carriage by sea, air, road and rail. The act prohibits the carriage of animals which are likely to give birth in transit. There are also provisions for the humane destruction of animals which are seriously injured during the journey. For further information covering the legal aspects of the transport of laboratory animals consult the list at the end of this chapter.

Containers

The range of containers available includes cardboard, plastic, metal and wood (Fig 3.1).

Cardboard containers are cheap, light and disposable but are not suitable for long journeys because the animal may escape by gnawing its way out, or the box may disintegrate from the action of urine or rough handling.

Plastic moulded containers are strong and sometimes autoclaveable but are expensive. Boxes made from folded corrugated plastic are cheaper and are quite widely used. Only strong plastic containers are accepted by the International Air Transport Association (IATA) for transporting rodents.

Metal containers have the advantage of being autoclaveable, reusable and are suitable for long journeys but provide little insulation and are expensive.

Wooden containers are becoming more expensive also and their use

3.1. *Animal transport boxes (a) corrugated plastic filter box (b) cardboard rabbit box (c) fibre board rodent box*

is gradually being restricted to transporting larger animals such as dogs and primates. Wood is very difficult to clean and eventually disintegrates through washing and sterilising.

Ventilation

Although containers should be draught-proof, ventilation must be provided, preferably with vents high on the sides of the box. If the vents are placed in the lid, the area in the box should be increased by 50% to allow better circulation of air. If boxes are going to be stacked together, the sides should be sloped so that an air space is created. When boxes with vents in the lid are stacked, spacer bars on the top or bottom of the boxes are incorporated to ensure adequate air circulation. The vents should be made of metal gauze small enough to prevent animals escaping or becoming trapped.

Specified pathogen free (SPF) animals must be transported in containers fitted with air filters. The pore size of the filter ranges from 0.5 – 3.0 μm and the filters are usually made of fibreglass.

Bedding

Absorbent bedding materials such as sawdust, woodchips and peatmoss are acceptable for most species. Rabbits and guinea-pigs may require hay in addition to the bedding as food or for comfort during the journey.

Food and water

All animals must be fed and watered before their departure – a dehydrated animal will suffer on a long journey. Water bottles should not be attached to boxes because movement will cause them to leak. Sliced apples, root vegetables or a wet mash will provide some moisture for long journeys; dry pelleted diet may be given in addition to fruit and vegetables.

Density

Animals should be packed with plenty of room for adequate movement and air circulation.

Labels

Labels must be firmly fixed to boxes and should contain the following minimum information:

1 Consignor's name, address and telephone number
2 Consignee's name, address and telephone number
3 A prominent 'LIVE ANIMALS' label
4 A 'THIS WAY UP' label
5 Name of consignee's railway station or airport from which they are to be collected.

If the animals are dispatched by air, the consignor should contact the consignee well in advance and advise him of flight numbers, and the date and expected time of arrival, to minimise delays at the airport.

Exporting laboratory animals

Animals must be in prime condition before they are exported, as strong, healthy animals will be better able to endure the stress of long journeys. The container for the journey must be of adequate size for the number of animals and must provide ventilation and some thermal insulation.

Before the animals are dispatched, a veterinary surgeon approved by the Ministry of Agriculture, Fisheries and Food must inspect them to ensure they are healthy and free of any obvious signs of diseases. Three signed copies of the veterinary certificate are issued as proof of the health status of the animals, one for the consignor and two to accompany the animals in transit, one of which will be required by Customs Officers, the other by the consignee.

If you are unfamiliar with exporting animals it is advisable to consult a shipping agent to handle the cargo and to process the necessary documentation. The agent will collect the consignment and deal with all the transport concerned to ensure the journey through customs is not delayed. If animals require attention during their journey, specific instructions must accompany them to enable a trained attendant to carry out duties such as feeding and watering.

On arrival at the destination the consignee should quarantine the animals until a satisfactory period has elapsed in which to confirm the animals are free of disease.

For information regarding the export of animals the following organisations should be contacted:

(a) Ministry of Agriculture, Fisheries and Food
(b) Embassy (in London) of country to which animals are being exported
(c) Shipping agent
(d) International Air Travel Association
(e) H.M. Customs & Excise

Importing laboratory animals

Most mammals entering the United Kingdom come within the scope of The *Rabies (Importation of Dogs, Cats and other Mammals) Order 1974* and have a minimum quarantine period of six months.

Imported animals may only be landed at specified airports or docks and have to be transported to the receiving laboratory by a licensed carrier in a suitable vehicle which has been approved for the purpose by the Ministry of Agriculture, Fisheries and Food. The receiving laboratory must also be licensed to hold imported animals; the licence is issued by the local divisional veterinary office of the Ministry of Agriculture, Fisheries and Food which will send a veterinary surgeon to inspect the premises regularly.

Recommended containers for laboratory animals

	Cardboard	Plastic	Wood	Metal
Mice	X	X		X
Rats	X	X		X
Hamsters			X	X
Gerbils	X	X		X
Chicks	X			
Guinea-pigs	X	X		
Rabbits	X		X	
Ferrets			X	
Cats			X	
Dogs			X	
Primates (small)			X	
Primate (>25 kg)				X

Documents authorising the acceptance of animals from abroad and their movement to licensed premises are obtained from the local Ministry of Agriculture, Fisheries and Food Divisional Veterinary Office.

Orders covering the transport of animals

1 Conveyance of Live Poultry Order 1919
2 Transit of Animals Order 1927 (cattle, sheep, goats, pigs)
3 Animals (Sea Transport) Order 1930
4 Transit of Animals (Amendment) Order 1931 (cattle etc.)
5 Transit of Horses Order 1951
6 Horses (Sea Transport) Order 1952
7 Live Poultry (Movement Records) Order 1952
8 Calves (Feeding in Transit) Order 1956
9 Live Poultry (Restrictions) Orders 1957–63
10 Movement of Animals (second) Order 1960
11 Transit of Calves Order 1963
12 Transit of Animals (General) Order 1973

Available from HMSO

Experimental animals and the law 4

All painful experiments on living vertebrates are controlled by law under the *Cruelty to Animals Act of 1876*. This act came into force as a result of an escalation of scientific work using living animals as research tools during the last century. There was concern amongst scientists and laymen alike that unnecessary cruelty could be inflicted on experimental animals. Also the discovery of anaesthetics led scientists to fear that living animals might be used for pseudo-scientific demonstrations that would be little more than party tricks. A growing body of opinion developed calling for some formal legislation to control animal experiments. T H Huxley and Charles Darwin were among the notable scientists adding weight to these developments.

In 1875 a Royal Commission was set up to investigate the whole question of animal experiments, under the chairmanship of Viscount Cardwell, the Secretary of War. From the recommendations of this Committee a Bill was drawn up and passed by Parliament in 1876 thus becoming the statute by which animal experiments are controlled today.

Thirty years later the government appointed a new Royal Commission under the eventual chairmanship of A J Ram. The Ram Commission decided that there was no need for new legislation but made several recommendations which were published in 1912. Among the most important of these were: that the number of Inspectors should be increased, that research should as far as possible be centralised and most important of all, that several changes be made to the 'pain condition' resulting in the condition that is written on licences today.

Application of the Cruelty to Animals Act

Before deciding if a procedure comes within the scope of the Act it should be remembered that the Act not only protects animals against unnecessary cruelty, but also protects the experimenter; any licensee conducting experiments on animals as laid down in the 1876 Act cannot be prosecuted under any other animal welfare act by any member of the general public.

To be within the Act a procedure must be all of the following: **an experiment, calculated to cause pain on a living vertebrate animal** (other than man). These terms are not defined in the Act but are clarified in 'Guidance Notes on the Law Relating to Experiments on Animals', published by the Research Defence Society, London.

An **experiment** is a procedure, the outcome of which is unknown in advance, the animal being used to provide an answer to a question.

A procedure is **calculated to cause pain** if it is liable to interfere to a material degree with an animal's health, comfort or integrity. Pain is deemed to include, in addition to noxious stimuli that cause distress, any adverse change in the animal's metabolism, any decrement in health or any interference in its reasonable comfort and well-being.

An animal is regarded as **living** if its heart is beating, it is breathing spontaneously and if its cerebral hemispheres and basal ganglia are intact. With the current use of artificial pacemakers and the techniques for sustaining life emphasis is put on the last point.

There are several restrictions imposed by the Act on experiments calculated to cause pain, the most important being:

1. The experiment must be performed with a view to the advancement by new discovery of physiological knowledge or of knowledge that will be useful in saving or prolonging life or alleviating suffering.
2. The experiment must be carried out by a person holding a Home Office Licence.
3. The animal must, throughout the experiment, be under the influence of an anaesthetic of sufficient power to prevent it feeling pain.
4. The animal must be destroyed at the end of the experiment before the effect of the anaesthetic has ceased. Section 3 restriction 4 of the Act states that if some serious injury has been inflicted or if pain will continue after the effect of the anaesthetic has ceased the animal should be killed while still under the influence of the anaesthetic.
5. Except in special cases, experiments must not be carried out to illustrate lectures.
6. No experiment may be carried out to acquire or attain manual skill.

Some of these restrictions may be lifted under the authority of the appropriate Certificate. The Certificates require in addition to a licence are summarised in the table below and should be read in conjunction with the conditions listed on the licence which are also reproduced in this chapter.

Licences and Certificates are legal documents. They are personal to the holder and delegation of authority under them is forbidden. The Home Office has given some guidance on the question of delegation.

Procedure	Horses, Asses Mules (Equidae)	Dogs & Cats	All other vertebrates
Under anaesthesia without recovery	Licence + Certificate F	L	L
Under anaesthesia with recovery	L + B + F	L + B + EE	L + B
No anaesthesia employed	L + A + F	L + A + E	L + A
Lecture or demonstration under anaesthesia without recovery	L + C + F	L + C	L + C

1. There is no delegation where two workers each with a licence and appropriate certificates carry out conjointly the operative or other procedures involved.
2. A licensee may permit unlicensed personnel to administer anaesthetics.
3. A licensee may allow someone to carry out simple mechanical duties such as restraint of animals, holding retractors or feeding experimental diets.
4. Subject to the above the prohibition of delegation is absolute.

Printed on pages 2–4 of every Home Office licence there are at least ten conditions which must be observed by the licensee.

Condition 1. Specifies the places where the licensee can carry out animal experiments. These are normally registered premises, approved by the Home Office as suitable for animal experimentation.

Condition 2. No experiment under any certificate may be performed until the licensee has been notified that the certificate has 'not been disallowed'. The submission of a duly signed certificate is not immediately followed by its coming into effect.

Condition 3. In all experiments under Certificate A or B (whether or not accompanied by Certificate E, EE or F), the following conditions are to be observed:

(a) If an animal at any time during any of the said experiments is found to be suffering pain which is either severe or is likely to endure and if the main result of the experiment has been attained the animal shall forthwith be painlessly killed.
(b) If an animal at any time during any of the said experiments is found to be suffering severe pain which is likely to endure such an animal shall forthwith be painlessly killed.
(c) If an animal appears to an Inspector to be suffering considerable pain and if such Inspector directs such animal to be destroyed it shall forthwith be painlessly killed.

These parts of condition three are known collectively as the *pain condition* or *pain clause*.

Condition 4. No operative procedure more severe than simple inoculation or superficial venesection may be adopted in any of the said experiments when covered by Certificate A.

Condition 5. For experiments under Certificate B an anaesthetic of sufficient power to prevent the animal from feeling pain must be used. Also strict aseptic precautions should be taken and if these fail and pain results, the animal shall be killed immediately whilst under anaesthetic.

Condition 6. Under Certificate C the animal shall be killed at the end of the demonstrations by or in the presence of the licensee.

Condition 7. Prohibits the use of curare or similar neuromuscular blockers unless permission has been obtained from the Secretary of State.

Condition 8. Licensees must keep written records of all experiments which must be available to the Home Office Inspector. At the end of each year details of experiments must be reported on a form provided for the purpose.

Condition 9. The Secretary of State should be sent a reprint of all articles describing experiments performed under a Home Office Licence by the licensee involved. The paragraphs describing the experimental procedure should be marked in the margin with details of the date and place of performance and the certificate(s) involved.

Condition 9a. Cinematograph film or video-tapes of licensed experiments must not be made without the written consent of the Secretary of State.

Other conditions that are sometimes added may require the work to be supervised by a senior licensee (this is not delegation) and may specify some procedures which must not be used such as the use of the eyes for withdrawal of blood or for the administration of substances. There may also be a condition prohibiting work under licence alone.

Annual Returns

Each year every licensee is required to fill in an Annual Return form and send it to the Home Office before 15th January. The return form has been recently revised in order to gain more information from the licensees about their experiments including a comprehensive breakdown of the type of work and species of animal used.

Plurality of experiments

With the present return form it is only necessary for the principal licensee to give detailed information about conjoint experiments. The other licensees simply have to report that they have carried out conjoint experiments. In the case of an experiment where a person performs some surgery on an animal to be used by another licensee, the surgeon carrying out work under Certificate B is regarded as the principal licensee. The licensee using the animal will be carrying out a conjoint experiment with the principal licensee, the surgeon. If the same animal is used for another experiment by another licensee using the same surgical preparation, that licensee will record his work as a separate experiment under Certificate A.

Normal animals used as controls to experiments are not considered experimental (if they are not subjected to pain as defined earlier) and should not be recorded on the Annual Return.

Application for a licence

It is important to decide if a licence is really needed. Any queries about this are best answered by the Home Office Inspector. The following procedures do not come within the scope of the 1876 Act:

Preparation of antisera, although testing the effect of serum to find appropriate dose levels would come within the Act.

Experiments on isolated tissue.

Preference trials (foods).

Comparison of weight etc, between animals fed on *wholesome* diets.

Simple behavioural experiments.

If a licence is required an application form must be obtained from HMSO and when completed must be signed by a Professor in a branch of medical science and by a President of one of various named Societies or Colleges e.g. the President of the Royal Society or the President of the Royal College of Surgeons. The Home Secretary has the power to disallow or suspend certificates and the Secretary of State grants licences and allows certificates on the advice of the Home Office Inspector. The *Guidance Notes on the Law Relating to Experiments on Animals* published by the Research Defence Society gives many examples of how licence applications should be worded. It is important that applicants for a licence describe accurately the work they propose to do.

In some cases the Home Office may require a senior licensee to give an *undertaking* that he/she will make him/herself responsible for the proper observance by the applicant of the provisions of the licence. Licences are normally granted for 5 years after which they must be renewed.

Registration of premises

The first condition of every licence is that all animal experiments must be carried out at specified premises registered for that purpose. The Home Office offers some guidelines and makes several recommendations about the suitability of premises for this work. One important requirement is that animals on experiment should not be on view to the general public either directly or indirectly through a window. Apart from common sense recommendations about animal welfare, the Home Office stipulates what information is required on an experimental animal cage label. The label must indicate (a) the nature (a brief description) and serial number of the experiment (b) the name of the principal licensee (c) the date when the experiment began and (d) if the cage holds more than one animal, the markings or other means of identifying the animal. For experimental designs involving large numbers of cages this information need not be duplicated on every cage. Most of the information can be given on a single sheet showing a cage plan which can be pinned up on the wall of the animal room.

When new animal units are being planned it is most important and indeed time-saving to involve the Home Office in the early stages of planning to make sure that the premises will be suitable for registration when completed.

Animals under experiment must not be removed from the registered premises without notifying the Home Office. The licensee may not carry out experiments on premises other than those specified on the licence; if he/she wishes to do so permission must be obtained from the Home Office.

Farm animals

There are some special considerations regarding the 1876 Act that should be borne in mind when dealing with farm animals. All experiments on equidae need to be covered by Certificate F. Cows, goats, sheep and pigs come under the classification of 'other vertebrates'. Normally only animals under Certificate A may be sold for slaughter and used for human consumption. Sometimes animals used under Certificate B may be kept long enough to be considered normal and, depending on the approval of the Home Office Inspector, may be sold for slaughter.

The *Medicines Act of 1968* states that if any drug other than for routine veterinary therapy has been administered to animals that are destined for human consumption or when their products are to be used likewise, a Test Certificate must be obtained from the Ministry of Agriculture, Fisheries and Food. The Test Certificate usually states that an animal may be slaughtered or its products used 20 days after administration of the drug has ceased.

With respect to the use of farm animals for raising antisera, if hormones, glandular extracts or sera of animal origin are to be used, permission must be granted by the Ministry of Agriculture, Fisheries and Food under the *Foot and Mouth Diseases (Sera and Glandular Products) Order 1939*. Most commercial products will have been approved and a statement to this effect will be on the bottle or vial label. Technically the normal raising of antisera is not considered an experiment under the 1876 Act as the outcome of the procedure is known; however, careful consideration should be given to whether a procedure is best regarded as an experiment to avoid any risk of possible prosecution under some other Act, for example the *Protection of Animals Act 1911*.

When working with farm animals one should be acquainted with the Acts concerned with various farm animal diseases. Under the *Diseases of Animals Act 1950*, in the event of an outbreak of any of the following, the authorities must be notified: anthrax, brucellosis melitensis (Ruminants, Swine, Equidae, RSE) cattle plague (R and S), epizootic lymphangitis (E), foot and mouth disease (R and S), Newcastle disease (poultry), glanders (E), parasitic mange (E), pleuro-pneumonia (Bovidae, B) rabies, sheep pox, sheep scab, swine fever (S), tuberculosis of the udder (SB), swine vesicular disease, epizootic bovine leukosis (B).

Sheep scab is again a problem in this country and at the time of writing it is compulsory in some areas to dip all sheep annually at a time specified (each year) by the Ministry of Agriculture, Fisheries and Food usually September – November.

Rabies is another notifiable disease which has become more of a risk to the UK in recent years as it has spread across Europe. The *Importation of Mammals Order 1974* restricts the importation of mammals and imposes stringent quarantine regulations. When using laboratory animals from abroad one should be acquainted with these regulations.

The *Movement of Animals (Records) Order of 1960 with Amending Order 1961* requires that a written record of all cloven hooved farm animals brought onto premises or moved from them is kept and is available for inspection by Ministry of Agriculture, Fisheries and Food Inspectors.

Other Acts

Other Acts that may also apply are: *The Dogs Act of 1906* which makes a number of provisions for the control of dogs and for dealing with strays. Of importance to the research worker is the clause forbidding the handing-over of stray dogs for experimental purposes. The *Protection of Animals Act 1911* makes it an offence to beat, kick, overdrive, ill-treat or cause necessary suffering to domestic or captive animals. This Act protects animals from cruelty by members of the public. If a research worker had decided that because the procedure he was to carry out was a marginal case and therefore need not be carried out under licence he would then be liable to prosecution under the 1911 Act should anyone object to his work.

The Protection of Animals (Anaesthesia) Act 1954 makes it an offence to carry out surgery of any kind on an animal without the use of a suitable anaesthetic, the 1876 Act lays down stricter controls of this aspect. *The Veterinary Surgeons Act 1966* makes it an offence for anyone other than a qualified veterinary surgeon to carry out veterinary procedures on an animal. Exceptions are medical practitioners or dental surgeons who may carry out treatment if so requested by a veterinary surgeon, and the owner of an animal or his employee. Procedures under the 1876 Act are also excepted.

The future

The 1876 Act is over 100 years old and it is a tribute to those who were responsible for the original draft that it has been flexible enough to cope with many of the changes in animal experimentation. However, there are some parts of the Act that require changing to fit modern needs and thinking.

In 1965 a report reviewing the whole question of animal experimentation was produced by a select committee under the chairmanship of Sir Sidney Littlewood. The Littlewood Report contains many recommendations for improving the existing legislation but unfortunately parliamentary time has never been afforded to consider these recommendations, although many have been accepted in principle and some are incorporated in the Home Office guidelines.

The principal recommendations were:

Less wastage of animals by avoiding duplication of research thereby reducing the numbers of animals used.
The procedure for signature of approval for licence applications to be replaced by a system of sponsorship.
The first licence issued should be probationary.
Experimental work giving rise to pain, lasting discomfort or incapacity should have prior approval of the Home Office.
There should be specific regulations governing the care and accommodation of laboratory animals.
The Home Office should encourage the professional development, qualification and training of animal technicians.
The complement of Home Office Inspectors should be increased to 21 and organised on a regional basis so that medical and veterinary qualifications are available in each region.
The Secretary of State should be given wide powers to make regulations to keep statutory control abreast of scientific developments.
There should be an advisory committee to advise on proposed experiments of a novel or controversial character.
The Advisory Committee should consider what information should be presented to the public.
Laboratory animal breeding establishments should be registered and inspected by the Home Office.

Recently, increased legislation imposed by the American Food and Drug Administration (FDA) and the Environmental Protection Agency (EPA) has resulted in changes in animal experimentation in the United Kingdom. It is anticipated that in time all animal experimentation will be carried out in accordance with these requirements which affect all aspects including diet and cage manufacture. At the time of writing it seems likely that the Council of Europe will produce a legislative document for member countries and future legislation in the United Kingdom will be in accordance with this.

Alternatives to animal experimentation

Much of the argument against the use of animals for research is based on the belief that there are satisfactory alternatives. Whilst undoubtedly techniques such as tissue culture and immunoassay do much to reduce the number of animals required for some aspects of research, at the time of writing there are no methods that would entirely replace the use of animals for research; expert opinion suggests there may never be such a time.

Euthanasia 5

Euthanasia is the humane killing of animals. There are various methods for most laboratory animals and the one chosen depends on a number of factors. Whichever method is used a number of points must always be observed.

1 Make sure that the animal really must be killed, for example because of ill health, surplus to requirement, or for experimental reasons.
2 Always try to act in a normal way, never giving the animal cause to be unduly frightened.
3 Never kill an animal within sight of others.
4 Never allow live animals to come within smelling range of blood; take particular care not to handle animals with blood on your hands.
5 Never use a method you cannot carry out efficiently and humanely.
6 Never use a physical method until you have practised on a dead animal and gained confidence with the procedure.
7 Always treat the animal with respect, never forgetting that it is a sensitive creature capable of experiencing fear if not rational thought.
8 Always make sure the animal is dead before disposing of the carcase. This can be confirmed by a lack of reflex, no respiration and some sign of urination due to relaxation of bladder muscles and rigor mortis.

Animals can be humanely killed by a number of methods the choice depending on the species concerned, the skill of the technician and the equipment available.

Dislocation of the cervical (neck) vertebrae is effective as this will break the spinal cord and will kill the animal quickly because neural control of vital organs such as the heart and lungs will immediately stop. Cervical dislocation is usually restricted to the smaller species and will be described for each species.

Decapitation is possible but is an unpleasant method and should be confined to special circumstances when it is the only suitable method.

Shooting using a humane killer is usually used for the larger species.

Humane killers are either the captive-bolt type or the free bullet type (Fig 5.1). The latter may be dangerous to use where hard surfaces such as walls or floors may cause the bullet to ricochet if it passes through the animal. In an emergency a rifle, revolver or shotgun may be used but never an airgun.

A firearms certificate issued by the police is required for all these weapons except the shotgun for which a shotgun certificate is necessary. With these methods the intention is for the bolt or bullet to severely damage the brain causing immediate unconsciousness and haemorrhage which will weaken and stop the heart thus causing death; the process can be accelerated by cutting a major blood vessel, for example the jugular vein.

A lethal gas may be used, usually in conjunction with a lethal chamber. Carbon dioxide, chloroform, ether and nitrogen can all be used in this way.

The most convenient way to administer a gas is from a pressurised cylinder through a reducing valve into the lethal chamber which should have an air outlet tap. Carbon dioxide (CO_2) is heavier than air and therefore the air outlet should be at the top of the chamber. An observation window is usually provided in the top. When killing small animals it is an advantage if the chamber is large enough to accept a

5.1. *Captive-bolt humane killer with blank .22 cartridge*

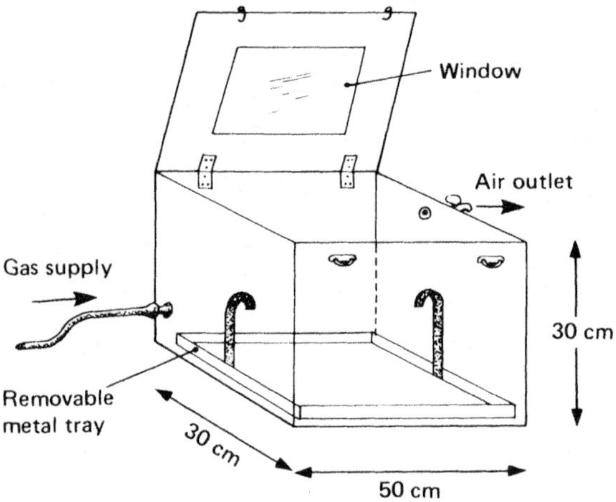

5.2. *Gas chamber suitable for killing animals up to the size of a rabbit*

standard cage (Fig 5.2). CO_2 is a safe gas to use, it is non-combustible, non-toxic and can be supplied in a pressurised cylinder. Chloroform or ether can be used by soaking a pad which is placed under a false mesh floor in the lethal chamber. The animals should not come into direct contact with the pad or the liquid. A desiccator makes a useful lethal chamber for ether or chloroform. In all cases the animals should be allowed to breathe a mixture of air and gas until they are unconscious. If using CO_2 or nitrogen from a cylinder the concentration of gas can then be increased. Once satisfied that the chamber is fully charged with gas the supply can be turned off. The animals should be left for 10–15 minutes before removal and should never be disposed of until it is certain that they are dead (note 8 page 42). Care should be taken when using ether because it is explosive and can be dangerous if high concentrations occur. Chloroform may cause sterility in mouse colonies and should therefore never be used in the animal room. It is also toxic and should only be used in well ventilated rooms.

An overdose of barbiturates may be used for most animals but very large amounts would be required for the larger farm species. Proprietary solutions of sodium pentobarbitone are available under various trade names e.g. Euthatal, Anasleep Forte, Sagatal, Nembutal. The lethal dose of a barbiturate is three times the anaesthetic dose. One source quotes a lethal dose of 100 mg/kg which gives a good safety margin. Administration by the intravenous route is the most suitable though the intraperitoneal and intracardiac routes may be used. Sodium pentobarbitone is highly soluble and a lethal solution may be made at a concentration of 1–2 g powder per 1 ml of water. The solution may be stored up to one week in a cool dark place. When choosing a method of

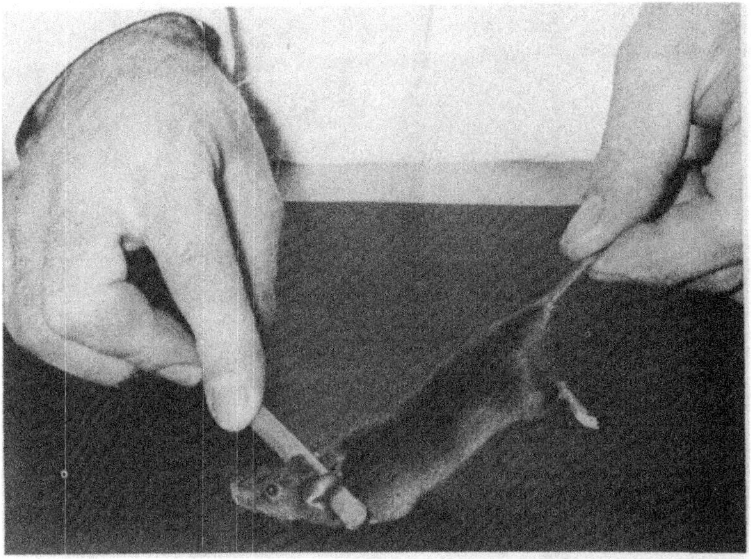

5.3. *Position for killing mouse by cervical fracture using a pencil*

euthanasia it must be remembered that barbiturates are relatively expensive.

Recommended methods

Mouse

- **lethal chamber.**
- **cervical dislocation.** Place the mouse on a solid surface, place a pencil across the back of the neck and with a swift movement pull back and up with the tail and push down with the pencil (Fig 5.3.) Dislocation of the neck breaks the spinal cord causing instant death: always check that dislocation has been achieved.
- **barbiturates.**

Rat

- **lethal chamber.**
- **cervical dislocation.** Hold the rat around its body with two hands, bring the rat sharply down striking the back of its neck against the edge of a sink or similar hard surface.
- **barbiturates.**

Guinea -pig

- **lethal chamber.**

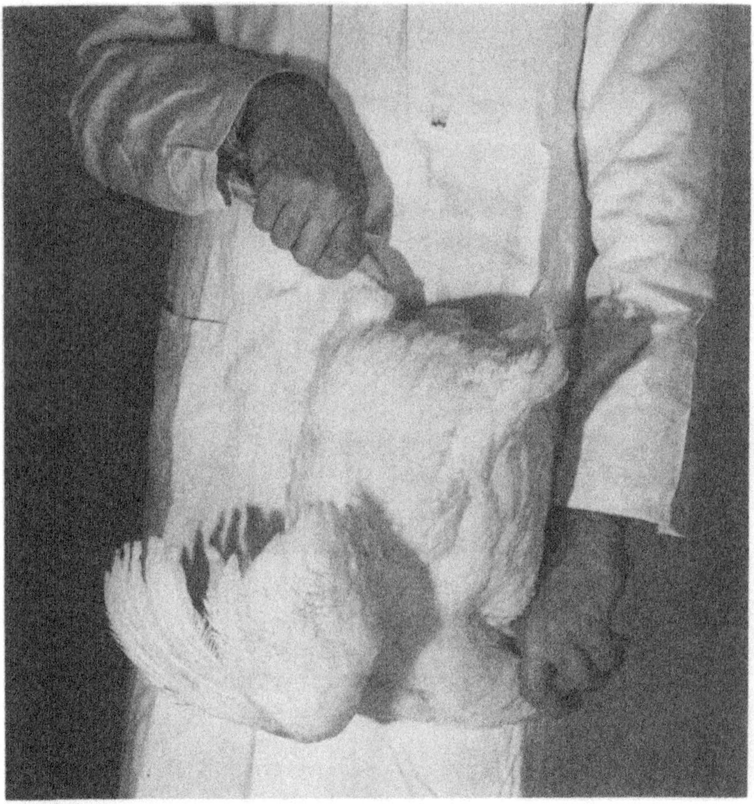

5.4. *Restraining position for wringing a chicken's neck*

– **cervical dislocation.** As for rat. An alternative method is to grasp the guinea-pig around its neck and to bring the animal sharply down still holding the neck firmly, thus performing a whip cracking movement which will break the guinea-pig's neck. The method must be practised on a dead animal.

Hamster
– as for rat

Rabbit
– **lethal chamber.**
– **cervical dislocation.** Hold by the scruff above the shoulders so that the head hangs down, hit behind the ears across the back of the neck with a heavy blunt instrument; with practice some people can use the edge of the hand (rabbit punch).

Alternatively hold the rabbit by its back legs in one hand and the head in the other, pull the head downwards and back until the neck breaks. This method is difficult with large rabbits.
- **barbiturates.**

Poultry
- **lethal chamber.**
- **cervical dislocation.** 'Wring' the neck as described for the rabbit (Fig 5.4). Alternatively the neck can be broken by using Burdizzo bloodless castrating forceps.

Reptiles
a sharp blow to the head with a blunt instrument or decapitation with a sharp instrument.

Fish
a sharp blow to the head.

Cat
- **lethal chamber.**
- **humane killer.** Hold by the scruff of the neck with the head on the ground. It is often convenient to hold the cat in a bag with its head out. The gun or humane killer should be aimed behind the ear towards the opposite eye (Fig 5.5).

5.5. *Aiming points when using a captive -bolt humane killer*

- **barbiturates.** The preferred method for the cat.

Dog
- **humane killer.** Reassure, feed if necessary, muzzle if it is fractious, steady its head, aim at a point made by diagonal lines crossing from ears to eyes and aim towards the spine (Fig 5.5).
- **barbiturates.** The preferred method for the dog.

Large animals including farm animals
- **humane killer.** The aiming points are illustrated in Fig 5.5. With a large animal such as the cow, care must be taken to keep out of the way when it falls.

Feeding and Nutrition 6

Wild animals can be classified according to the type of food they eat: herbivores eat vegetable material, carnivores eat flesh and omnivores eat a mixed diet. When living in the laboratory most of the commonly used species become omnivorous and are fed diets made to standard formulae for each species. Standardised diets based on cereal supplemented with good quality protein, vitamins and minerals have come into general use for most species including primates and are often intended to be complete, though fruit is often added to diet to provide extra vitamin C and help to prevent boredom.

Food is the raw material from which all animal tissues are built – new growth in the young animal and replacement in the adult. Minerals form the skeleton and proteins build the cells that form the soft tissues of the body. Food also provides most of the energy needed for warmth and movement and for the complex biochemical reactions which take place in the body. Vitamins and some minerals act as catalysts for many reactions and must be provided in the diet. Since these reactions take place in an aqueous medium water is vital although it is not considered to be a nutrient in the normal sense of the word. Diets usually contain about 10% water and water for drinking is usually supplied to animals from water bottles or automatic drinkers.

Food intake will rise during growth when young animals are forming new tissues, during gestation when the female is providing for the developing fetuses, and during lactation when the female is producing milk for her young or if the animal is very active or the environment cold. Animals are usually fed *ad libitum* but may be given a restricted diet to curb body weight (for example obese mice 5 g of food per day). When animals are fed *ad libitum* they will generally eat until their energy requirement is satisfied. All other nutrients in the diet must be related to the energy content of the whole diet in order that enough of these nutrients are eaten in the correct proportion.

Food is required to produce and maintain animals, therefore the composition and requirements of the animal with respect to proteins, carbohydrates, fats, minerals and vitamins should be reflected in the composition of the diet. A balanced diet is one which supplies all the nutrients required by the animal in the correct proportion for growth and maintenance and without unreasonable excess of any one constituent.

Proteins

Proteins are composed of carbon, hydrogen, oxygen, nitrogen, sulphur and phosphorus (CHONSP). Proteins are broken down into their component amino acids during digestion and it is from these amino acid 'building blocks' that new animal protein tissues are made. Each tissue has a different amino acid pattern but the body as a whole requires these amino acids, of which there are about twenty, in a particular pattern. If one amino acid is present at a low level in the diet, that is, the diet is deficient in it, when it has been used up by the body none of the other amino acids can be utilised and they are wasted. (The one which is deficient is called the 'limiting amino acid'.) Some amino acids can be made in the body but others, the essential amino acids, must be provided in the diet. Dietary protein must possess amino acids in the correct ratio; animal or 'first class' protein, for example fish-meal, meatmeal, is superior, but vegetable or 'second class' protein, for example soyabean meal, groundnut meal, is often deficient in the amino acid lysine. Animal protein is expensive and is often mixed with cheaper vegetable protein to provide the correct balance of amino acids, synthetic amino acids may also be added notably lysine and methionine. Cereals contain about 10% protein (the germ) and between one third and one half of the total protein in the diet may be provided by cereals. If too much protein is included in a diet excess amino acids are deaminated in the liver and the nitrogenous amino part is excreted and wasted, the remainder (CHO) being used to provide energy. This is uneconomic as the expensive protein is merely used instead of carbohydrate which could be provided at a much lower cost.

Carbohydrates

These provide much of the energy needed by the animal, they are composed of carbon, hydrogen and oxygen (CHO). They are broken down into simple sugars and oxidised (burnt) in the tissues to provide energy. The liver converts excess sugar into glycogen for short term storage or into fat for long term storage for later use when the body's energy requirement exceeds that provided by the diet. Some fibre and cellulose is classified as carbohydrate but can only be efficiently utilised by ruminants (for example cows and sheep) or animals with an active caecum (for example guinea-pigs and rabbits). Fibre is necessary for all animals as roughage to help the peristaltic action of the gut.

Fats and oils

These are also composed of carbon, hydrogen and oxygen but have less oxygen and are higher in energy having twice as many joules per gramme as cereals or protein. They are incorporated into diets to provide extra energy and as a source of essential fatty acids. They form a vehicle for the transport of the fat soluble vitamins (A, D, E and K) and increase palatability and pelleting qualities.

Minerals

Inorganic elements are included in diets in varying amounts; when present in minute quantities they are known as trace elements. Their functions are wide ranging.

Calcium and phosphorus (Ca and P). These constitute the largest proportion of minerals in the diet and form the major part of bones and teeth. Phosphorus is needed in many enzyme systems and calcium is an essential component of the blood clotting mechanism. The ratio of calcium to phosphorus in diets should be between 2:1 and 1:1. Vitamin D is essential for the efficient absorption of calcium from the intestine and its deposition in bone. If the diet of young animals is calcium deficient their bones do not form properly producing a condition known as rickets where the bones are slightly pliable and bend under the weight of the animal.

SOURCES: Calcium – limestone flour, calcium phosphate, sterilised bone meal, dried milk.
Phosphorus – cereals especially the by-product bran.

Magnesium (Mg). Constituent of bones and enzyme systems. It is not usually necessary to add magnesium to the diet, because it occurs naturally in other constituents.

Sodium, potassium and chlorine (Na, K and Cl). These three elements are involved in many enzyme systems, the functions of the blood and with maintenance of the correct amount of water in the body tissues (water balance). Again it is not necessary to add potassium to diets but common salt (sodium chloride, NaCl) is usually added. Salt also improves the palatability of the diet.

The following are trace elements, which are required in minute amounts.

Iron and copper (Fe and Cu). Iron forms the 'haem' part of haemoglobin, and copper is necessary for the production of haemoglobin.

Cobalt (Co). A component of vitamin B_{12} (cyanocobalamin).

Manganese (Mn). Necessary for enzyme and reproductive functions.

Zinc (Zn). Forms part of the respiratory enzyme carbonic anhydrase.

Iodine (I). Present in the hormone thyroxine produced by the thyroid gland; a deficiency of iodine leads to an enlargement of the thyroid known as a goitre. (Derbyshire has a very low level of iodine in the water and goitres in the human population were common, known as 'Derbyshire neck'.)

Vitamins

Vitamins act as catalysts which alter the speed of reaction of many of the metabolic processes that take place in the body. The retention of vitamins in the body varies considerably; some have to be supplied daily while others are stored for a considerable time. Vitamins are classified according to their solubility: vitamin C and the B complex are water soluble and A, D, E and K are fat soluble. The diet should contain a small percentage of fat in which the fat soluble vitamins may be dissolved and transported.

Vitamin A. Sometimes called the 'anti-infection' vitamin as a deficiency causes a breakdown of the epithelial tissues allowing infectious agents to enter the body. The skin, eyes and reproductive organs are often affected resulting in blindness and reproductive failure. Vitamin A is also a component of the pigment visual purple (rhodopsin) which is produced in the rods of the eye and enables the animal to adapt to poor light conditions. A deficiency of the vitamin results in 'night blindness'.

SOURCES: The precursor, carotene, occurs in green plant material, maize and carrots. Most animals are able to convert carotene into vitamin A in the body; an exception is the cat which must be fed pre-formed vitamin A. Liver and fish liver oils are excellent sources of vitamin A.

Vitamin D. Necessary for the deposition of calcium and phosphorus during the formation of bones in young animals, a deficiency results in rickets.

New World monkeys (for example marmosets and squirrel monkeys) and birds can only utilise D_3 whereas most animals are able to utilise D_2 or D_3. (Naturally occurring vitamin D contains equal amounts of D_2 and D_3.)

SOURCES: Vitamin D is synthesised in the skin when exposed to direct sunlight. Fish liver oils are rich in vitamin D_3. Synthetic preparations are also available.

Vitamin E. Necessary for reproduction, a deficiency results in abortion or resorption of the fetuses and infertility.

SOURCES: The germ of cereal grains is rich in vitamin E. Synthetic compounds are also available.

Vitamin K. An essential factor of the blood clotting mechanism, a deficiency may result in a fatal loss of blood from minor cuts.

SOURCES: Vitamin K is synthesised by micro-organisms in the lower gut. It is thought that animals which practise coprophagy obtain the vitamin this way. Germ-free or SPF animals may need their diet supplemented with synthetic vitamin K because their intestinal flora is absent or reduced.

The vitamin B complex. All the B vitamins are concerned with enzyme systems.

Vitamin		Function	Deficiency symptoms
B_1	Thiamine	Carbohydrate metabolism	
B_2	Riboflavin		'Curly toe' in chickens
	Nicotinic acid (Niacin)	Involved in growth rate	
	Pantothenic acid		Loss of hair pigment
B_6	Pyridoxine		Convulsions, dermatitis and anaemia
	Biotin		Paralysis
	Folic acid		Anaemia and poor growth
B_{12}	Cyanocobalamin		Poor reproduction

SOURCES: All are present in liver and all except B_{12} in cereals and yeast.

Vitamin C (Ascorbic acid). Vitamin C is necessary for the formation of collagen in connective tissue and the intercellular cement between endothelial cells of blood vessels. Guinea-pigs show classic signs of deficiency – scurvy, bleeding gums and swollen joints, all examples of the breakdown of tissue.

SOURCES: Most animals can synthesise their own ascorbic acid but primates (including Man), guinea-pigs, fruit bats and a bird (the Red-rumped Bul-Bul) are deficient in an enzyme needed to make ascorbic acid and are therefore dependent on a dietary source of the vitamin. Fresh fruit and vegetables are rich in vitamin C but synthetic ascorbic acid is readily available. The vitamin is rapidly destroyed by wet heat and contact with metals and is therefore best added to the diet rather than the drinking water. The vitamin cannot be stored by the body and therefore has to be given continually.

Water

The physical and chemical properties of water combine to make it the most important constituent of living matter. It is also the most abundant, comprising at least 70% of most soft tissue. An animal can withstand starvation for long periods but deprivation of water can be lethal in a very few days.

Water is the vehicle in which most of the other constituents of the body are dissolved or suspended. The products of digestion are transported in an aqueous medium from their site of absorption in the gastro-intestinal tract to the tissues and many of the consequent waste products are excreted in solution in the urine. Water plays an important role in heat regulation of the body; because of its high heat capacity it can absorb much of the heat generated during metabolism so that body temperature does not rise excessively. Heat is dissipated through the skin and in the water vapour exhaled from the lungs. Because of the physical properties of water (its high latent heat of evaporation) relatively large heat losses are achieved by the loss of a small volume of water.

Antibiotics

These are used commercially as growth promoters but may give rise to resistant strains of bacteria and so should not be fed routinely to animals. Guinea-pigs and ruminants which are dependent on their gut flora should not be fed antibiotics.

Foodstuffs 7

Manufacture and physical presentation

The vast majority of diets used in animal houses are purchased in a state ready to be fed to animals without further preparation. Only rarely, such as in nutrition laboratories or where large numbers of farm animals are used, are diets milled and compounded in the laboratory or associated feed mill.

Laboratory animal diets are usually produced according to a fixed formula, that is, prescribed amounts of a named raw material (as in the examples given in Appendix 1.5 p. 196). Alternatively they may be formulated to provide nutrients to a specified level, the amounts and type of ingredient being varied according to the nutritional analysis of each ingredient and its cost. Diets for commercial livestock are usually produced by this method. The choice of method of formulation for laboratory animals will be influenced by the type of experiment the animals are to be used in. If nutritional studies are involved the fixed formula method is probably to be preferred as the introduction of an alternative ingredient, groundnut meal instead of soyabean meal for instance, could interfere with the experimental result. The level of nutrients supplied by fixed formula diets will vary however according to the composition of the ingredients which are affected by the geographical regions and climates in which they were grown.

Some ingredients arrive at the mill ready ground to a coarse powder, soyabean meal, fishmeal and middlings for example, while others, notably the cereals, are usually ground on site. The bulk ingredients are weighed out, often automatically, and placed into the mixer which may have a capacity of several tonnes. The small items, vitamins, minerals and extra amino-acids are added to the main mix after having been pre-mixed with some of the bulk load to ensure even distribution.

When the meal has been thoroughly mixed it may be pelleted. Steam or molasses is added as a binding agent just before the meal is forced through a die consisting of a thick metal plate with holes in it, the extruded meal is broken off by a rotating blade to form pellets. These pellets are commonly 4 mm in diameter for rabbits, guinea-pigs and poultry, 10 mm in diameter for rodents and farm animals and much larger cross-sections, 40 mm × 15 mm ('jumbo-nuts') for farm livestock such as pigs. Some laboratory diets are commercially available coated with silicon dioxide which prevents the pellets sticking together should they be sterilised in an autoclave in the laboratory. A similar property may be imparted by dusting the pellets with powdered talc.

A new presentation of diet has appeared on the market in recent years: the expanded or pasteurised diet. This has been autoclaved during manufacture and has an aerated structure resembling a hard sponge. The advantages claimed for this diet are a lower microbial load and economy in feeding in that less wastage occurs as the pellets are gnawed. This saving is offset to some degree by the higher cost of the diet and it is also bulkier to store. Conventional pelleted diets may also have a reduced microbiological count as they became heated to about 60°C by the process of pelleting and by passing the pellets immediately through enclosed chillers rather than allowing them to cool in the open air, recontamination is reduced.

Handling and storage

The need to store diets is one of convenience and economic necessity, deliveries every few days are rarely practical unless very large quantities are involved. It is common policy to order no more diet than will be used within one month. Most animal houses have a main store where the bulk of the diet is held, when it is required in the animal room it is emptied from the sack and stored short term in small containers such as plastic skips or plastic dustbins on wheels. Most laboratory animal diets are packed in 3 ply paper sacks holding 25 kg. Diets for SPF units are usually packed in tough polythene bags which are packed into paper outer sacks or fibreboard boxes before being irradiated.

The methods of storage should aim to reduce to the minimum the physical and chemical deterioration of the diets. Very fresh pellets are more susceptible to physical damage than those a few days old and care in handling should therefore be exercised to prevent excessive amounts of dust being produced which will not be eaten by the animal. Sterilised diet must be handled with considerable care to prevent the polythene bag becoming punctured so rendering the contents non-sterile. Direct loading from the delivery lorry onto conventional or wheeled pallets will reduce the handling individual bags will receive as well as being labour saving. Diet should be stored on pallets or duckboards to allow air to circulate under the stacks; if placed directly on the floor the bags at the bottom are liable to 'sweat'. There should be a small gap between stacks and adjacent walls to allow air to circulate and to provide easy access.

Each delivery should be stacked separately and the oldest diet used first, new diet should never be placed on top of old. By common agreement most manufacturers stamp each bag with the date the diet was made but it is useful to place 'Use first' signs on the appropriate piles.

The food store should be dry, well ventilated and the temperature held at about 10–15°C (50–60°F) and not allowed to drop to the point where condensation will form (dew point), this will depend on the dampness (relative humidity) of the air. The chemical degradation of some vitamins and amino-acids will be accelerated at higher temperatures and moulds and fungii will thrive if conditions are damp and may cause considerable heating of the diet.

As direct sunlight can destroy certain vitamins, the food store should be windowless though with the use of paper sacks which themselves afford some protection, this is not essential. Sunlight may cause local heating and drying of diet, the latter reducing palatability.

The diet store should have tight-fitting doors and windows and ventilators should be screened to prevent the ingress of insects, rodents and other pests. It should be possible to seal the store so that its contents may be fumigated should infestation occur. *This should only be carried out by suitably trained persons owing to the hazardous nature of the chemicals used, for example ethylene oxide or ethylene dichloride. The store should be thoroughly ventilated after such treatment.* Although well-aired diet can be fed to animals it is best discarded if it was treated because it was infested.

The short term storage of diet in the animal room is a matter of convenience. Shallow, covered plastic skips on wheeled tables are preferable to dustbins, they are more easily cleaned, less dust is produced as the bag is emptied and less bending is needed to scoop out the diet when feeding. Both the food store and the animal rooms should be kept clear of spilt food which will attract rodents and other pests.

Pests

Insects and arachnids (for example spiders, mites) may be introduced into the store with the diet or may gain access through cracks and ill fitting doors and windows. They may cause damage by physically destroying grains or pellets, contaminating diet with their excreta or acting as vectors of disease.

The weevils and beetles (Order *Coleoptera*) are named according to the foodstuffs they infest – Grain weevil (*Colandra granaria*), rice weevil (*Colandra oryzae*), biscuit weevil or bread beetle (*Stegobium paniceum*) and bacon or larder beetle (*Dermestes lardarius*), although

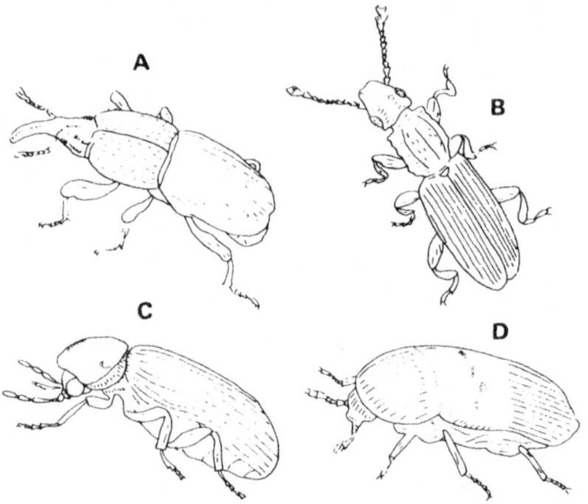

7.1. A *Grain weevil* (Sitophilus granarius) *4 mm*
B *Saw-toothed grain beetle* (Oryzaephilus surinamensis) *3 mm*
C *Biscuit weevil* (Stegobium paniceum) *4 mm*
D *Larder beetle* (Dermestes lardarius) *10 mm*

they are also found on other grains and meals. They cause damage by boring into grains to lay their eggs (Fig 7.1).

The presence of the mill or flour moth (*Ephestia kuhniella*) is characterised by tangled masses of web and meal dust often in the lids of bins or the corners of sacks.

Cockroaches, *Periplaneta americana* (American), *Blatta orientalis* (Oriental) and *Blattella germanica* (German) live in warm, damp, dark places such as heating or service ducts and emerge during darkness to feed on spilt foodstuff. Their presence may not be suspected until a darkened room is illuminated, especially out of working hours, when large numbers may be seen.

Mites (Order *Acarina*) may be introduced with the diet and are usually only a problem in diet stored for a few weeks that has become damp. Individual mites are microscopic but are usually seen as a fawn coloured dust when present in great numbers along with meal debris and their exreta on the floor at the base of hessian meal sacks or on the surface of meal stored in a bin. If this deposit is examined very closely it will appear to seethe, a line traced across the dust will disappear after a few minutes due to the movement of the mites.

Flies will cause little direct damage to stored foodstuffs but should be excluded as they may be the vectors of infection. Wild rodents may also present problems and a trap is probably the most effective way of dealing with them; it is unwise to use poisons near foodstuffs as the animals may die and decompose in inaccessible places.

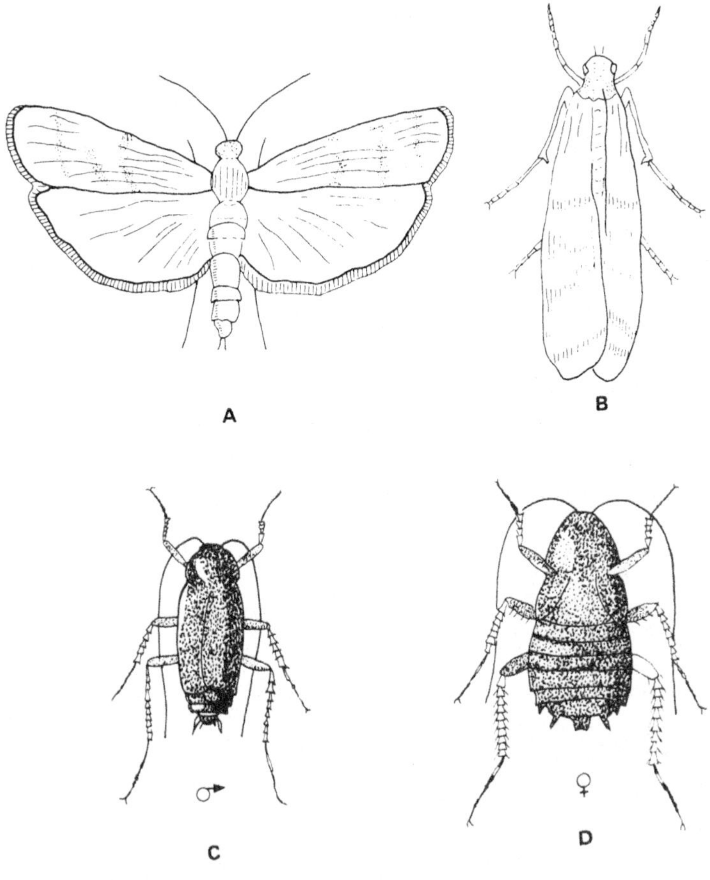

7.2 A *Mill moth* (Ephestia kuhniella) *10 mm*
B *Warehouse moth* (Ephestia elutella) *10 mm*
C, D *Common cockroach* (Blatta orientalis) *30 mm*

The Ministry of Agriculture, Fisheries & Food (MAFF) produces some excellent booklets on arachnid pests of foodstuffs; line illustrations of some of the common pests are shown above.

Breeding laboratory animals

8

In establishing and running a breeding colony some system must be adopted if only to decide which animals to put together in order to mate. Before selecting a system consideration must be given not only to the number of animals required but also to the needs of the experimenter, such as closely related animals (inbred), or animals carrying a specific characteristic (hairless, obese).

One should aim to use the highest quality of animals available, whether they have been bred in the unit or brought in from a disease free reliable source, as these animals will produce not only stock for experiment but also future breeders. Subsequent generations of animals will reflect the status of the initial colony if husbandry and management are efficient.

When setting up a breeding colony the following requirements must be considered in order to decide on the system to be used and the number of breeders required.

(a) Number to be produced per week, month or year
(b) Sex
(c) Age and weight at issue
(d) Specific requirement, for example inbred or random bred.

Breeding systems

Inbreeding

A strain is said to be inbred when it has been maintained by brother × sister mating for twenty consecutive generations; genetically the animals are alike (homozygous) to a degree where they are actually identical. Only about 1% of genetic difference remains. 100% homozygosity will never be achieved partly because of spontaneous changes that occur in the genetic material (mutations).

There are 2 systems of inbreeding, 'single line' and 'parallel line', though a combination of both can also be used.

In the single line (Fig 8.2a) system one pair of animals from each generation is mated, though to safeguard the survival of the colony

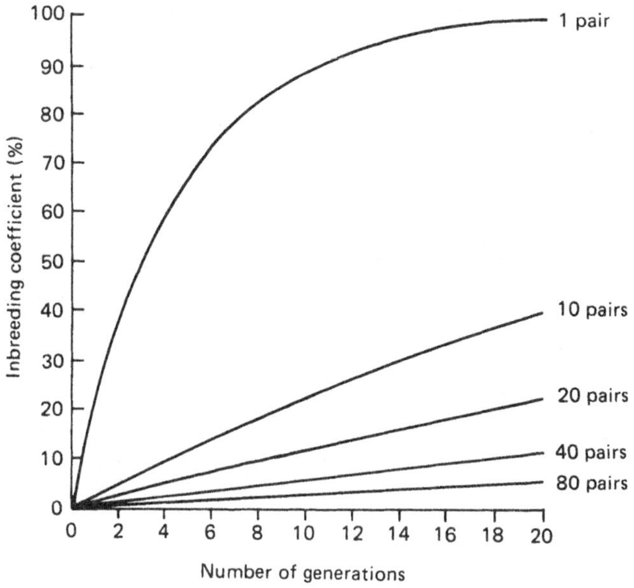

8.1. *Inbreeding coefficient*

more than one pair should be selected but the next generation chosen from only one pair; offspring of the other pairs will go to stock. Parallel lines (Fig 8.2b) emerge when 4 or 5 pairs of animals are selected for breeding from one pair and are subsequently bred; 4 or 5 pairs of single lines will emerge and because these run parallel to each other they are called parallel lines. If this system is continued there is a possibility of genetic variation arising as a group of single line subcolonies is produced.

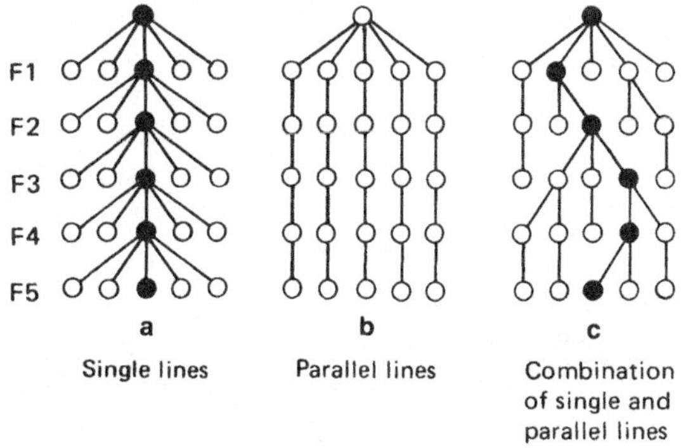

8.2. *Inbreeding systems (●, ○ represents 1 breeding pair)*

Often large numbers of inbred animals are required so a combination of both systems (Fig 8.2c) is used whereby one pair is mated to continue the colony line and other pairs mated to produce either breeders or stock for experiment. Lines formed by the latter pairs are usually halted after three generations as genetic variation may occur. In practice this can be carried out by using the 'traffic light' system (Fig 8.3). 1 inbred pair is given a white breeding record card, the offspring, or first generation are given green breeding record cards, the second generation are given yellow breeding record cards and the third generation are given red breeding record cards. All the offspring from the 'red' pairs are used for experiment, the red card being an indication to halt breeding (hence 'traffic light' system). To continue the line 1 pair is selected from the original white card pair, this pair will also be given a white card, their offspring will be given a green card and so on, allowing the continued production of animals.

After the first 5–6 generations of inbreeding 'inbreeding depression' may become obvious by lowered fertility, shorter life span and greater susceptibility to disease.

Random breeding

Random breeding is often confused with outbreeding or non-inbreeding. A true random breeding system occurs where animals are given numbers and are paired with the aid of a random numbers table. However most colonies that are random bred are the result of the

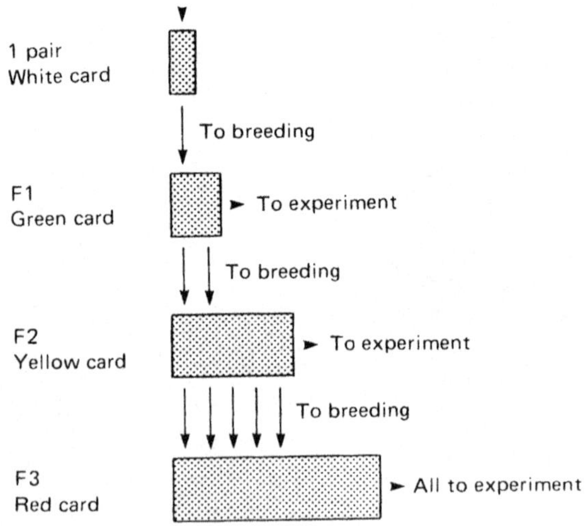

8.3. *'Traffic light' breeding system*

selection of suitable healthy stock (page 64) that are paired with little or no knowledge of their parentage.

Outbreeding or non -inbreeding

An outbred or non-inbred system of breeding must lead to as much genetic diversity as possible. Unrelated animals are mated to provide vigorous healthy stock which are genetically unalike (heterozygous). It is therefore important to maintain records identifying breeding stock to ensure that related animals are not paired. In a closed colony – where no other animals from an outside source are introduced – it may eventually be impossible to avoid pairing related animals: at this stage only those most distantly related should be paired.

Mating systems

Monogamous pairs

A male and female are paired when sexually mature and remain together throughout their breeding life thus taking advantage of the post partum oestrus (when mating will take place immediately after the birth of the young). This system requires one cage per breeding pair so will require more space and labour than trios or harems, it is however virtually the only system for inbreeding and may also be necessary where precise breeding records are required.

Trios

As the name suggests trios are groups of 3 animals kept together throughout their breeding life, in the ratio of 1♂ : 2 ♀. Before using this system one must ensure the type of cage used is adequate to house the 3 adults plus 2 litters of growing animals. The advantages are that the animals may be inbred or random bred and one extra female is housed per box thus housing more animals in fewer boxes. A disadvantage may arise if both females produce young at the same time as it will not be known which offspring belong to which female.

Harems

Groups of animals are housed together in large numbers depending on the size of the animal and the cage. The ratio of males to females is usually 1:5; if too many males are housed together fighting may occur and poor productivity will result. The females become pregnant and either litter down in the harem cage or are removed to a littering cage; the stage of pregnancy is determined by palpation, or by the general appearance of the female. When the litter has been weaned from a separated female she is returned to the same harem for re-mating. In species that exhibit a post partum oestrus there is no possibility of mating occurring when the female is removed from the harem cage to a littering cage; therefore production time is lost. This system is labour and space saving.

Selection of breeding stock

The object of selecting breeding stock is to consistently produce robust, healthy and fertile animals. By using rigid criteria as a means of selecting breeding stock one can have confidence in the animals' ability to withstand abnormal changes in the environment such as heating failure, fluctuation in temperature and some diseases.

Criteria for selection of breeding stock

(a) The female parent should have a good overall breeding record and should not be prone to killing her young.
(b) The animals should show no signs of ill health, deformities or other oddities.
(c) If possible animals should not be selected from the first or second litters as they are usually small and do not reflect the size and standard of future litters.
(d) Select animals from litters showing:
    average litter size for the strain
    average weaning weight for the strain
    equal sex ratio within litters.
(e) The parent animals must not be aggressive towards the handlers.

Recording

The random system requires fewer records than the inbred or outbred systems where extensive records must be maintained and each animal identified with an ear mark or some other code to distinguish it from another in the colony. If an animal escapes and there is any doubt about its identity, to avoid mistakes such as putting the animal into the wrong cage and causing genetic contamination, it should be humanely killed.

Information for breeding colony records

(a) Female's identity, parentage and date of birth.
(b) Mating date and identity of male.
(c) Date of birth of litter, number and sex of young.
(d) Number of young stillborn and deaths before weaning.
(e) Date of weaning litter, number in litter and sex.
(f) Number of animals issued for experiment or breeding stock.
(g) Information at post mortem examination or cause of death.

In systems using monogamous pairs, trios or harems it will be necessary to include on the card the identity, parentage and date of birth of the male.

The record card may be used as a cage label or given a cage number and kept in a file drawer. Generally it is preferable to have one record card per female or per pair and all the information regarding breeding performance and health is put on that card alone. On the death of the female the card is removed from the cage or file and put into the 'deceased' breeder file where it can be matched up with litters and parents if cross reference is required in the future.

Reproduction

Whereas many plants and simple animals can reproduce asexually, mammals, fish, reptiles, amphibians and birds all breed by sexual reproduction. Spermatozoa from the male fertilize eggs (ova) from the female which implant onto the wall of the uterus. When fully developed the fetuses leave the uterus and pass into the vagina and are expelled when the female gives birth (parturition).

Male reproductive organs

The male reproductive organs are the penis, the testes or primary sex glands which produce spermatozoa, the epididymides and vasa deferentia. The prostate gland and Cowper's gland (accessory sex glands) produce fluids which combine with and serve as a medium to transport sperm during ejaculation. In the mouse, rat and guinea-pig it is the ejaculate which forms the vaginal plug (page 92).

Female reproductive organs

The female reproductive organs are the ovaries or primary sex glands, which produce ova, the fallopian tubes which lead the ova towards the uterus and where fertilisation takes place, the uterus in which the embryos develop and which is closed by a strong sphincter muscle, the cervix, and beyond that the vagina or birth canal where semen is deposited by the male during copulation. The outer opening of the vagina is the vulva.

Hormones

The primary sex glands or gonads, the testes and ovaries, secrete sex hormones which are important in the development of secondary sexual characteristics, and in the female are responsible for the cyclical changes that occur in the reproductive system. The gonads do this under the influence of secretions from the pituitary gland, the gonado-trophic (gonad altering) hormones. Of these, follicle-stimulating hormone (FSH) stimulates the growth of the eggs in the ovary and the ovarian follicles that surround them and induces the production of oestrogen by the ovary. The luteinising hormone (LH) induces ovulation which is followed by the development of the corpus luteum. Luteotrophic hormone (LTH), also secreted by the pituitary, causes the corpus luteum to secrete progesterone which results in sexual quiescence in the reproductive cycle.

Some mammals such as the rat, mouse, hamster, guinea-pig and monkey are known as polyoestrous (having many oestrous cycles). Because an increased release of LH occurs in waves determined by a fundamental physiological rhythm, causing ovulation, they are also known as spontaneous ovulators; ovulation occurs at regular intervals and is basically independent of the environment. The sheep and goat for example are also polyoestrous and spontaneous ovulators but are affected by the environment, day length and temperature, and only come into breeding condition in late autumn; they are known as seasonal breeders. Species such as the rabbit and ferret are known as induced or reflex ovulators. They have no spontaneous cycle, ovulation is brought about by the release of LH as a result of the stimulation of mating. The domesticated rabbit may mate throughout the year but the ferret is affected by the environment and is therefore a seasonal breeder.

The oestrous cycle

Different species have oestrous cycles of varying length, this is dependent on the hormones produced by the pituitary. In the guinea-pig, pig and man there is a spontaneous oestrous cycle with a luteal (sexual quiescent) phase of about 16, 21 and 28 days respectively. There is spontaneous production of FSH and LH followed by LTH which activates the corpus luteum which prepare the uterus to receive the embryo (luteal phase). If fertilisation does not take place the cycle starts again, but when embryos are present the activity of the corpora lutea is extended throughout pregnancy. In the rat and mouse there is spontaneous production of FSH and LH leading to a 4 to 5 day oestrous cycle coinciding with maturation of ovarian follicles and release of ova, there is negligible production of LTH so the corpora lutea are practically inactive. After mating the nervous stimulation lifts the inhibition of LTH production, the spontaneous cycle stops, corpora lutea are formed, become active and maintain pregnancy. In the rabbit and ferret there is no spontaneous cycle. The pituitary produces FSH only, causing follicular growth (and therefore production of oestrogen by the ovary giving rise to oestrus) only. On the stimulus of mating LH is produced thus allowing the ovary to shed ova and form corpora lutea. Fertilisation and pregnancy can then take place.

In some species such as rat and mouse it may be necessary to detect oestrus for timed matings. The stage of the oestrous cycle may be determined by taking vaginal smears. If timed matings are required the animals should be paired at the pro-oestrous stage. Synchronisation of oestrus in the female is possible by using the Whitten effect which increases the proportion of mice in oestrus on a given day.

Whitten found that if female mice of breeding age were placed together in a cage the majority tended to become anoestrous[1]. If after a few days a male was put among them, either in the same cage or placed in a separate wire cage within sight and smell of the females, then a high proportion of the females would come into oestrus approximately three days later. Females can be stimulated by using vasectomised male mice, which will avoid fertile matings. It is also known that adult females effect the secretion of gonadotrophin in young male mice resulting in advanced puberty[2].

Vaginal smears

A simple and quick method of identifying the stage of oestrus is to make a vaginal smear. The animal (usually rat or mouse) is restrained and a small metal spatula or moistened cotton wool swab (Fig 8.4) is inserted gently into the vagina and rotated against the vaginal wall. The instrument is withdrawn and a smear made on a clean glass slide. The smear can be air dried or fixed more quickly by passing the underside of

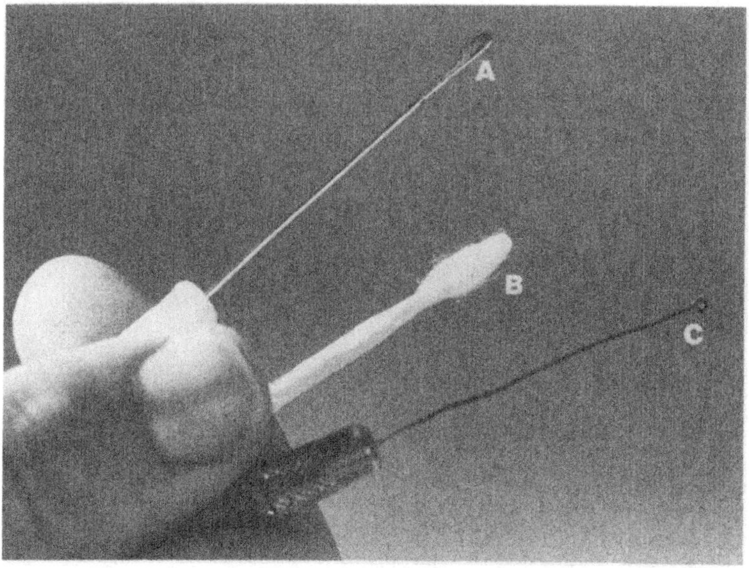

8.4. *Instruments for taking vaginal smears (a) stainless spatula (b) cotton bud (c) wire loop*

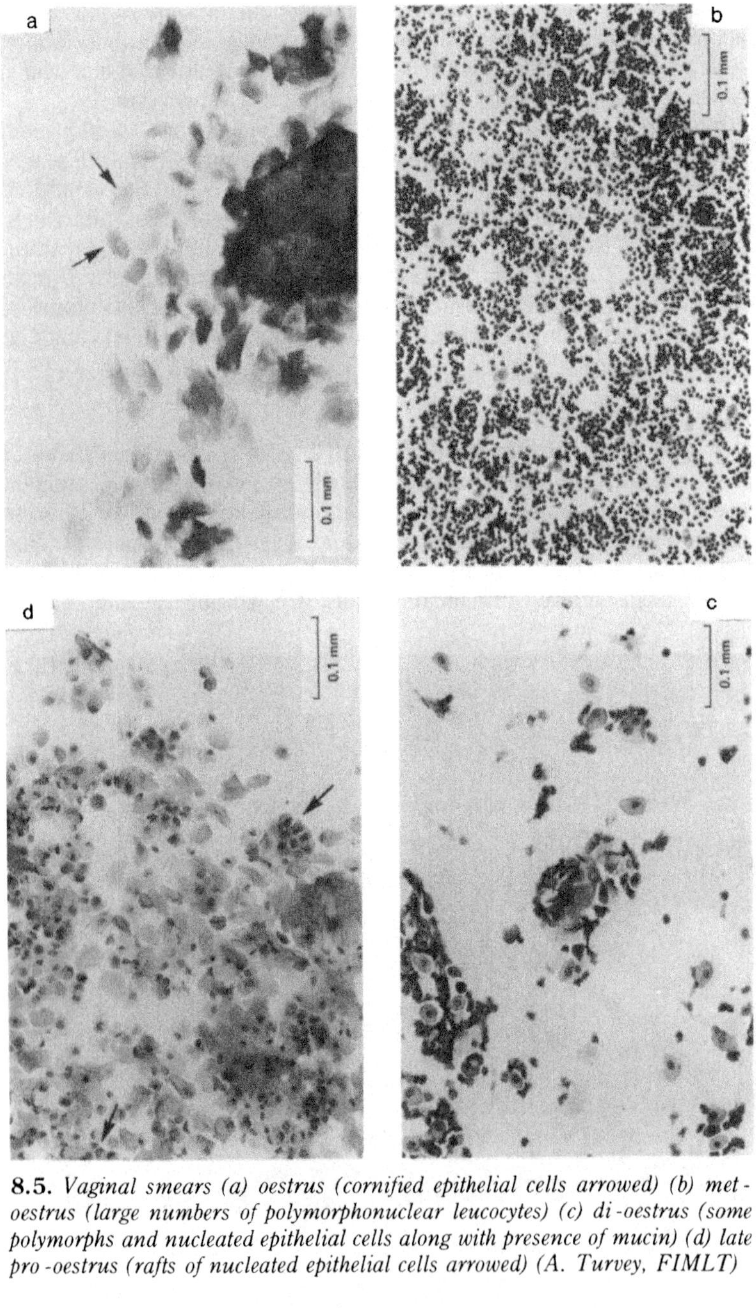

8.5. *Vaginal smears (a) oestrus (cornified epithelial cells arrowed) (b) met-oestrus (large numbers of polymorphonuclear leucocytes) (c) di-oestrus (some polymorphs and nucleated epithelial cells along with presence of mucin) (d) late pro-oestrus (rafts of nucleated epithelial cells arrowed) (A. Turvey, FIMLT)*

the slide rapidly through the flame of a bunsen burner. If it is necessary to stain the smear suitable stains are haematoxylin and eosin and/or methylene blue. When the smear is examined under the microscope the stage of the oestrous cycle may be identified. The oestrous smear contains only many large cornified epithelial cells with no visible nucleus; if mating has recently occurred spermatozoa will also be present. Metoestrus follows and a smear made in this phase will show a decreasing number of cornified epithelial cells and many leukocytes. This is followed by dioestrus – the smear will show many leukocytes and epithelial cells and often some mucus. Pro-oestrus follows dioestrus – the smear will show only nucleated epithelial cells which tend to occur in rafts. It is at this stage that animals should be paired for mating.

References

1 Whitten, W K (1958) Modification of the oestrous cycle of the mouse by external stimuli associated with the male. Changes in the oestrous cycle determined by vaginal smears. *Journal of Endocrinology*, **17**, 307–313.

2 Pheromones and mammalian reproduction. Advances in Reproductive Physiology. A McLaren (Ed.) 1966 155–177. Academic Press. New York.

Manipulations and Special Techniques 9

The usual routes for injection of substances are subcutaneous (under the skin), intradermal (into the skin), intraperitoneal (into the peritoneum or abdominal cavity), intravenous (into the vein) and intramuscular (into a muscle).

Mouse, rat and hamster

Injections

Subcutaneous injections are made under the loose skin of the neck (Fig 9.1).

Intraperitoneal injections are made through the lower part of the abdominal wall into the peritoneal (abdominal) cavity, slightly to one side of the mid-line to avoid piercing the bladder. The loose skin on the neck should be firmly grasped between the thumb and index finger, and for the rat and mouse the tail is held between the third and fourth finger of the same hand. The animal should be held with the abdomen upper-

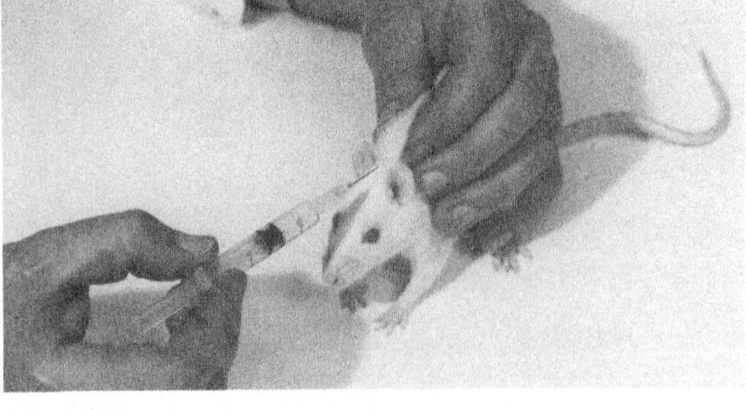

9.1. *Restraining a rat for a subcutaneous injection*

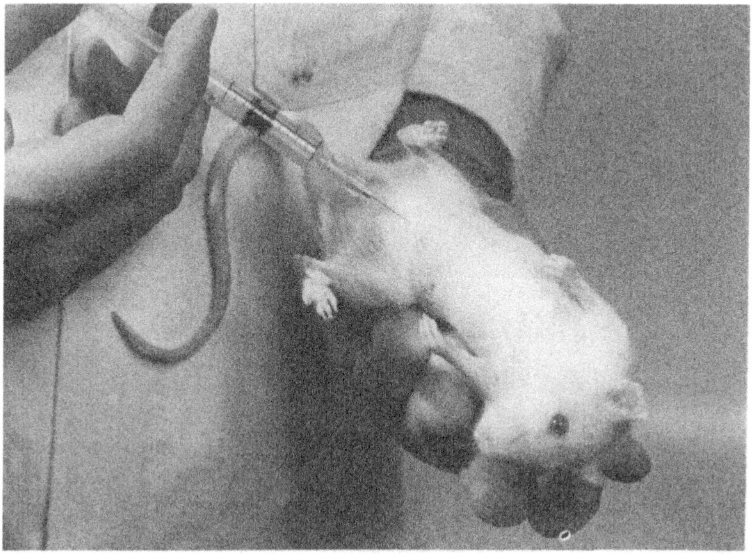

9.2. *Restraining a rat for a intraperitoneal injection*

most and with the head tilted slightly downwards so that the gut slips towards the diaphragm (Fig 9.2). The skin around the abdominal area should be taut before proceeding with the injection.

Intravenous injections are usually made into the lateral vein of the tail, near the tip (Fig 9.3). The animal should be warmed beforehand by immersing the tail in warm water to dilate the vein, or by warming the animal near a lamp, with the exception of albino animals. These have more delicate eyes which may be damaged by bright light. For the hamster, which does not have a suitable tail, the saphenous or cephalic vein must be used.

Intramuscular injections are best made into the posterior thigh muscle, the animal should be restrained using the method described under intraperitoneal injections. An assistant may be necessary to restrain a rat when performing intramuscular injections.

Collection of blood

Small blood samples may be obtained from the tail vein in the rat and mouse by nicking the tail with a scalpel blade after both have been cleansed with an antiseptic. The animal must be kept warm under a lamp to encourage a good flow of blood. Other possible sites include the orbital venous plexus of the eye and the femoral vein; these techniques, however, require considerable skill and should not be attempted by

9.3. *Using a restraining cone for rat tail vein injection*

inexperienced staff. (A prohibition condition may be imposed on Home Office licences.) Larger samples of blood may be withdrawn by syringe from the heart. The animal should be anaesthetised, placed on its back and the heart located under the 5th to 6th rib by palpation with the index finger.

Oral dosing

Substances may be given orally:

(a) Mouse – hold by the loose skin of the neck with thumb and index finger restraining the tail.
(b) Hamster – with the 3rd and 4th fingers hold by the loose skin of the neck, in the palm of the hand.
(c) Rat – hold with the palm of the hand over the animal's back with the thumb and index finger around and below the lower jaw.

The substance may be given in a syringe fitted with a measured length of semi-rigid plastic or a short needle with a bulb soldered on to the end to prevent damage to the oesophagus (Fig 9.4). The length of the tubing is critical for different species if damage to the stomach is to be avoided. The dosing needle is lubricated with corn oil or liquid paraffin and inserted into the mouth over the tongue and pushed gently towards the oesophagus. From the oesophagus it should pass easily into the stomach, but if resistance is felt it should be withdrawn and another attempt should be made.

Guinea-pig

Injections

Subcutaneous injections are made under the loose skin of the neck parallel with the underlying muscle. An assistant must hold the guinea-pig's shoulders firmly with one hand whilst restraining the lower part of its body with the other, so that the injection site is exposed.

Intraperitoneal injections require the guinea-pig to be held on its back with the abdomen uppermost. The needle is inserted slightly to the right or left of the mid-line away from the bladder into the peritoneal cavity.

Intravenous injections are difficult. The marginal ear vein may be used; if the ear is slightly warmed the vein is reasonably accessible. Alternatively the femoral vein in the leg can be exposed with the guinea-pig suitably anaesthetised. With care the resulting lesion will require only minimal suturing.

Intramuscular injections are made in the posterior thigh muscle.

Intradermal injections are made into a shaved site which has been cleansed with an antiseptic. A needle (26 gauge by 10 mm long) is inserted at an angle into but not under the skin to its full length and withdrawn slowly as the fluid is injected.

Collection of blood

Small quantities of blood may be obtained from the ear vein. Larger samples must be taken from the heart of an anaesthetised animal or from the femoral vein of a restrained animal.

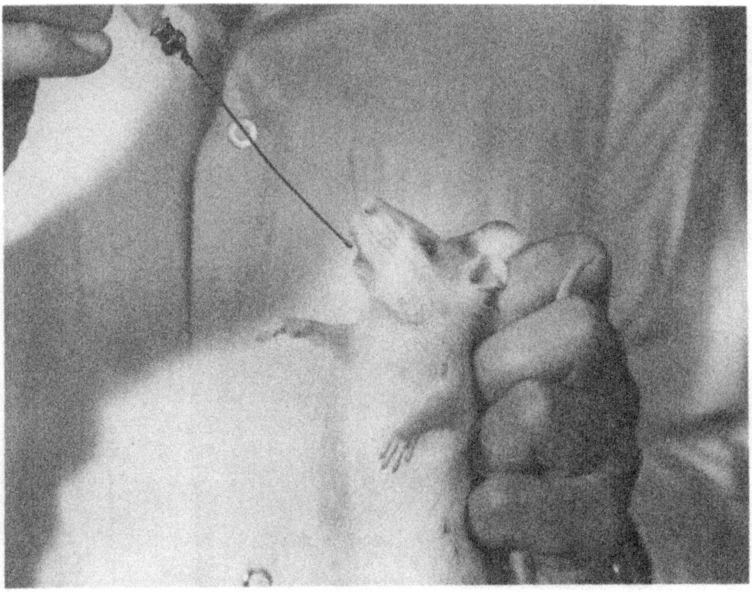

9.4. *Restraining a rat for oral dosing*

Oral dosing

Guinea pigs are not good subjects for oral dosing as they always seem to regurgitate copious amounts of food material when dosing is attempted. This increases the risk of material passing down the trachea into the lungs.

Rabbit

For most manipulations rabbits can be restrained in purpose-made boxes or they can be wrapped in a cloth.

Injections

Subcutaneous injections are usually made under the loose skin of the neck or flank. An experienced person may do this single handed, an inexperienced person may require an assistant to restrain the animal.

Intraperitoneal injections are made into the lower abdomen. The animal is held by an assistant; the loose skin of the neck should be grasped firmly with one hand whilst the other hand supports and restrains the hind limbs as it is turned abdomen uppermost. The method is safer if a table or bench can be used to support the rabbit while it is being restrained because of the danger of breaking its back.

Intravenous injections are made into the marginal ear vein more easily if it has been dilated by warming beforehand (Fig 9.5).

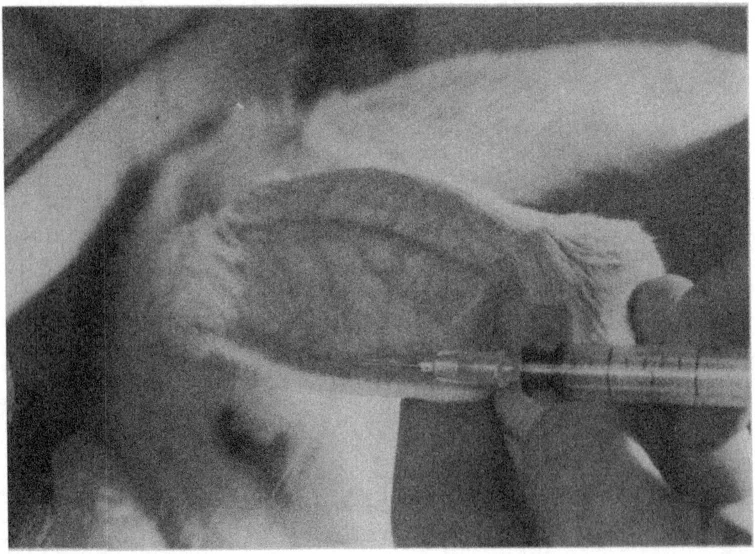

9.5. *Injecting into the marginal ear vein of a rabbit*

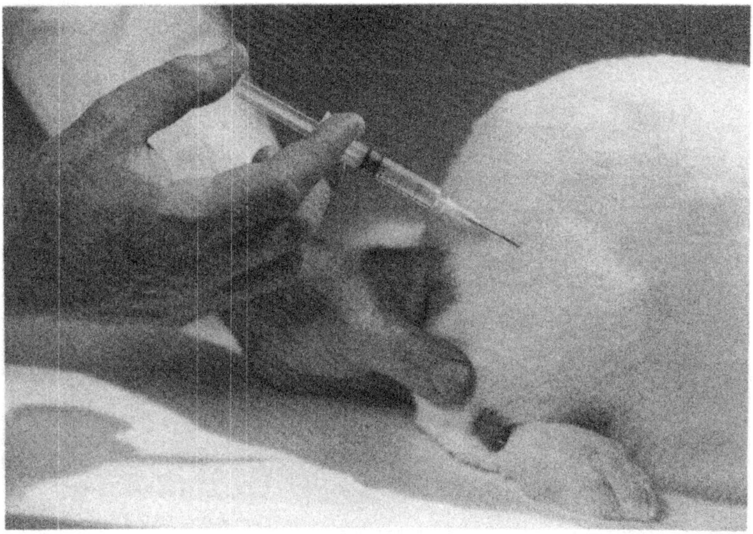

9.6. *Intramuscular injection in the rabbit*

Intramuscular injections are made into the posterior thigh muscle (Fig 9.6). The rabbit must be held by an assistant.

Collection of blood

Blood samples may be taken from the marginal vein in the ear or by cardiac puncture. Collection from the ear vein necessitates the area being shaved and cleansed with antiseptic before the vein is nicked with a scalpel and the blood is collected in a syringe or container. Cotton wool should be available to stem the flow of blood when the procedure has been completed. The artery can also be successfully used as a site (Fig 9.7). Larger quantities of blood may be taken from the heart of an anaesthetised rabbit lying in a dorsal position.

Oral dosing

Dosing is usually performed orally using a hypodermic syringe with a bulb soldered on to the end of a long needle. The technique is similar to that described earlier for the rat.

Cat

Injections

Subcutaneous injections are made under the loose skin of the neck. If small quantities are to be injected, the cat may be held under the left

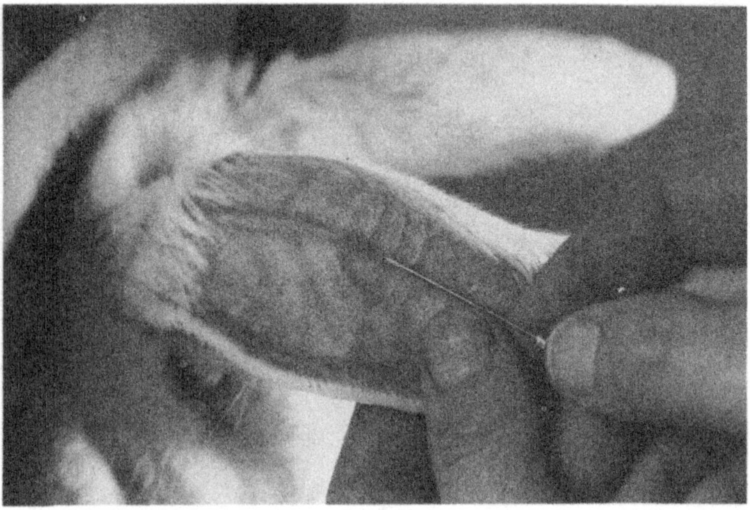

9.7. *Position for collecting blood from the ear artery of a rabbit*

arm, with its body pressed against the operator to prevent it struggling. When injecting large quantities an assistant should restrain the cat.

Intraperitoneal injections are performed with the cat standing in a normal position. The assistant grasps the loose skin at the neck with the left hand to restrain the head and places the right arm around the body. The needle is inserted through the skin in the mid-line of the abdomen into the peritoneal cavity about 5 cm tailwards from the caudal end of the sternum.[1]

Intravenous injection sites are identical to those used for withdrawal of blood.

Intramuscular injections are made into the posterior thigh muscle.

Collection of blood

Blood samples may be collected from the cephalic vein which lies between and above the radius and ulna, the saphenous vein which lies between and above the tibia and fibula or the jugular vein in the neck. The withdrawal site is clipped, shaved and cleansed with an antiseptic. Pressure is applied to the vein below the bleeding site to dilate it and the needle is gently pushed into the vein and blood is withdrawn into the syringe. 'Vacutainers' are available for obtaining given quantities of blood. The containers have had the air removed and when the double-ended needle is inserted into a vein and the stopper is pierced, the desired quantity of blood is drawn into the sealed tube. This is a very quick and efficient method of sampling blood, particularly if many samples are required.

Oral dosing

Substances are given in capsule form or mixed with milk or food. To dose orally, restrain the animal under the left arm, pressing it to the operator's body, open the mouth by pulling the head back and holding the jaw open, place the tablet onto the rear of the tongue, close the mouth and massage the throat until the tablet has been swallowed. It is advisable to check this a short time after administration. Another method of restraining a cat for oral dosing is to wrap it securely in a strong piece of cloth such as a laboratory coat, to prevent it scratching the operator or causing injury to itself.

Dog

Injections

Subcutaneous injections are usually made in the neck.

Intraperitoneal injections are made after the skin has been clipped, shaved and cleansed with antiseptic. The dog is restrained on its back and a 19 gauge by 40 mm needle is passed through the skin into the abdomen and into the peritoneal cavity.

Intravenous injections are identical to the method used for the collection of blood.

Intramuscular injections are made into the thigh approximately half way down the limb. The substance should be injected slowly to avoid pain.

Collection of blood

Blood samples may be obtained from the cephalic, saphenous and jugular veins. To take a blood sample from the cephalic vein the area of the site is clipped, shaved and cleansed with antiseptic, the vein is dilated by constricting the blood flow above the site (that is to say towards the heart) ready for the needle to be inserted. There is no need to keep the vein raised while taking the sample.

Dosing

Oral dosing using a stomach tube is a straightforward technique if the dog is accustomed to the procedure. It may be necessary to sedate an untrained animal before passing the stomach tube.

Liquids may be poured into a pouch of the lower lip, the nose is held slightly upwards and the throat massaged.

Tablets may be placed on the back of the tongue with the jaws open. Massaging the throat when the jaws are closed will help the dog to swallow the tablet.

Monkey

Injections

Monkeys must be sedated for the following procedures.

Subcutaneous injections are made under the loose skin of the neck.

Intraperitoneal injections are performed after the skin has been clipped, shaved and cleansed with an antiseptic. The needle is passed through the abdomen wall and into the peritoneal cavity, having pulled the skin away from the organs lying beneath.

Intravenous injections are made into the femoral vein. Pressure should be applied to the vein to dilate it before the injection is administered.

Intramuscular injections – as for the dog.

Cages are available with a movable partition (crush cage) which restrains the monkey at the front of the cage and removes the need to handle the unanaesthetised animal for injections.

Collection of blood

Blood may be collected from the femoral vein. The monkey should be placed on its back with its feet and arms secured and suitably restrained. The pulse of the femoral artery should be detected, the femoral vein is situated slightly medial to the artery. A 19 gauge by 32 mm needle is introduced into the vein and blood is withdrawn slowly. Care must be taken not to puncture the femoral artery, as this will result in heavy bleeding.

Dosing

Oral dosing with a polythene tube is a common method for administering liquids. Care must be taken not to damage the trachea, oesophagus or the stomach while the fluid is administered slowly through a tube with a large syringe.

Ferret

Injections

Subcutaneous injections are made under the loose skin of the neck and can be made without an assistant restraining the animal.

Intraperitoneal injections are made through the abdomen wall into the peritoneal cavity. The neck and shoulders are firmly grasped with the free hand and the hind limbs are held between the operator and the bench.

Intravenous injection – no superficial vein is suitable.

Intramuscular injections are made into the posterior thigh muscle.

Collection of blood

Blood may be withdrawn from the tail vein or the heart. For cardiac punctures, anaesthesia with a 50% air / 50% CO_2 mixture has been found satisfactory. The animal is laid on its back and a 19 gauge by 50 mm needle is inserted just off the mid-line below the sternum and passed under the ribs, through the diaphragm into the heart. This method is easier than the more conventional method of passing the needle between the ribs because the intercostal muscles are exceptionally tough. After withdrawing the needle and removing the anaesthetic mask recovery is extremely rapid.

If a few drops of blood are required the last 5 cm of the tail should be shaved, and lightly smeared with Vaseline to prevent contamination of the blood sample. While the animal is firmly held, the vein is sharply but gently incised at an oblique angle with a narrow scalpel blade (Fig 9.8). The blood that wells from the vessel may be collected with a syringe or small capillary tube.

Dosing

Liquids may be administered orally while the ferret is held around the neck and shoulders in one hand. The dosing tube should be made of material that is non-chewable and that cannot be punctured by the ferret's teeth. The tube is introduced behind the teeth at the side of the mouth and the liquid emptied on to the back of the tongue. The ferret should have no difficulty swallowing the liquid when this method is used.

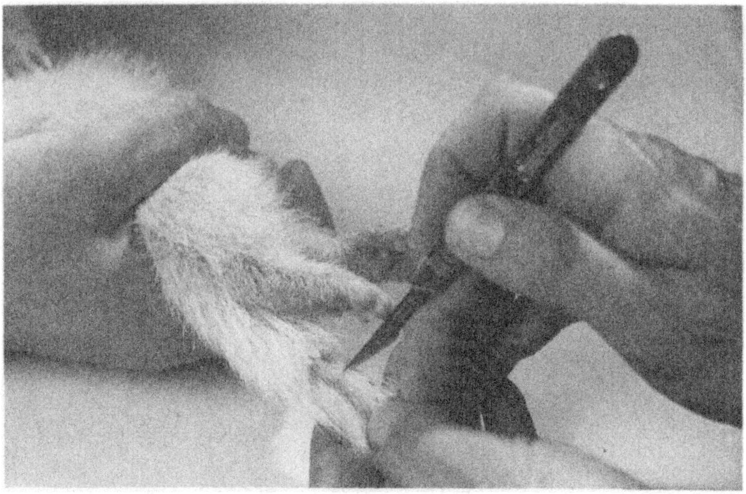

9.8. *Incising the tail vein of a ferret*

Nasal administration

As assistant holds the ferret in an upright position with the head tilted back while the operator sprays fluid from an inoculating sprayer into the nostril. The animal may sneeze after such treatment and there is some danger of infection being passed to the staff. Therefore, it is advisable to wear face masks, particularly if a virus solution is being administered.

Chicken

Injections

Intraperitoneal injections are made through the mid-line of the abdominal wall thus avoiding penetration of the abdominal air sacs. (Handling – see collection of blood.)

Intravenous injections are made into the wing vein. The feathers are removed from underneath the wing which exposes the vein to allow a 19 or 20 gauge × 17 mm needle to be inserted under the wing tendon towards the point of bifurcation of the wing vein (Fig 9.9).

Collection of blood

Blood samples may be collected from the wing vein or the heart. To collect a sample from the heart the bird is placed on its back on a table. An assistant holds the legs in one hand and with the other hand the neck is eased over the edge of the table. A 21 gauge × 38 mm needle attached to a syringe is inserted between the two clavicles slightly downwards with a little negative pressure being applied to the plunger.

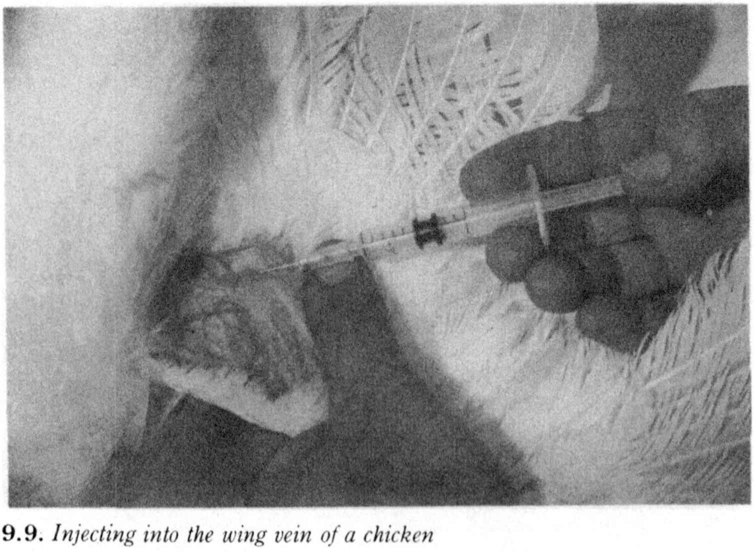

9.9. *Injecting into the wing vein of a chicken*

Immediately the needle enters the heart, blood will flow into the syringe.

The wing vein is another site used for collecting blood. This is a large well-exposed vein but considerable difficulty can be experienced when withdrawing blood from it. The bird is held in a dorsal position and the

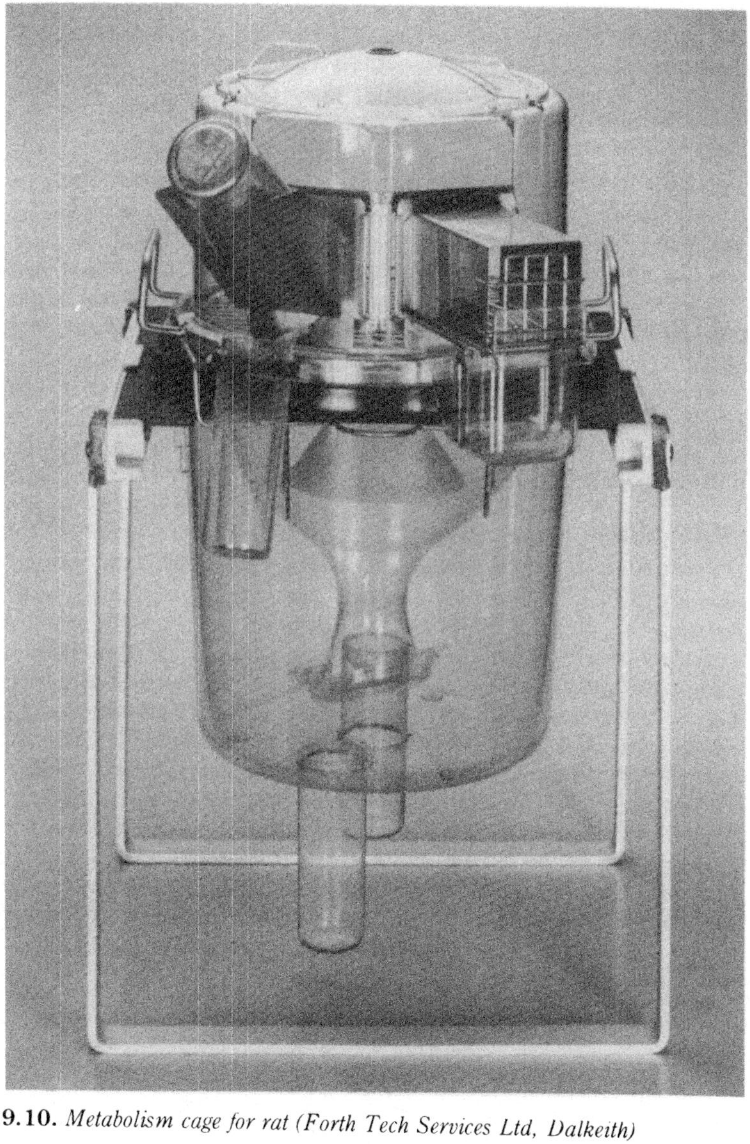

9.10. *Metabolism cage for rat (Forth Tech Services Ltd, Dalkeith)*

feathers are removed from underneath the wing, which is held in the operator's left hand. This exposes the vein and a 19 or 20 gauge × 17 mm needle is inserted under the wing tendon towards the point of bifurcation of the wing vein. This is visible as a well-defined area in the hollow of the wing. By entering the wing under the tendon it has been found that there is a better chance of penetrating the vein without damaging it. When the needle is clearly in the vein, provided excessive suction is not applied, blood may easily be withdrawn.

Collection of Urine and Faeces

Small samples of urine may be obtained from mice and rats by exerting slight pressure over the bladder, the urine expelled is collected in small tubes. If uncontaminated urine and faecal samples are required over a certain period, metabolism cages are generally used. Part of the cage consists of a wire mesh floor suspended over separators which channel the urine into one container and the faeces into another with minimal contamination of one with the other (Fig 9.10).

Metabolism cages are designed not only for the smaller species but for animals such as the dog, pig and sheep. If it is not important to have an uncontaminated faecal sample, it may be collected directly from the floor of the cage or the pen.

Taking Body Temperature

One of the most common methods of taking an animal's body temperature is by using a clinical thermometer, although the system of using a thermistor probe connected to a direct read-out display is a quick and efficient method which is gaining popularity. The clinical thermometer has a working range of 35–43°C (95–110°F) and can be obtained with a long or short bulb (Fig 9.11). The long bulb is used for smaller animals and the short bulb for animals such as the dog or monkey. A small kink prevents the mercury column dropping back when it has moved up to show a rise in temperature thus enabling readings to be taken away from the animal. The thermometer is re-set by shaking the mercury column down 12 mm (½″) below the lowest graduation on the scale and care must be taken not to warm the thermometer with the hands. Obviously a thermometer cannot be placed in an animal's mouth as in human patients and it is therefore placed through the anus into the

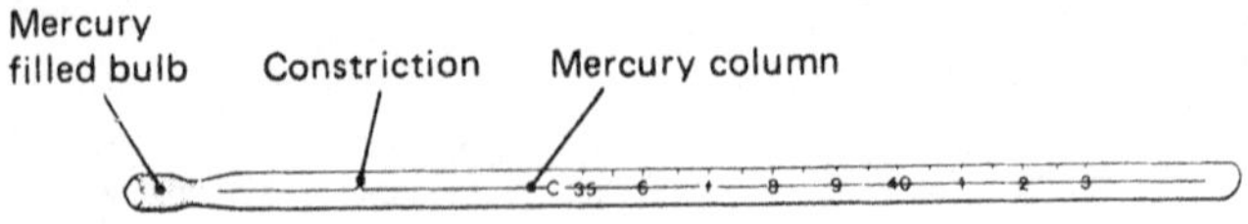

9.11. *Clinical thermometer*

rectum. For rectal temperatures the bulb and part of the stem are smeared with Vaseline or a similar lubricant. The animal is restrained in a comfortable, relaxed position so that it will not move or try to dislodge the thermometer. Once this resting state is achieved, the thermometer is gently inserted into the rectum until the whole of the bulb is inside. It is important to insert the thermometer to the same depth at each reading, otherwise the temperatures may vary considerably. If the thermometer meets some obstruction in the rectum, do not force it, withdraw and rotate it gently past any faecal matter to obtain a reading. After taking the temperature the thermometer should be cleaned and sterilised in 70% alcohol.

References

1 Scott, P P (1976) The UFAW Handbook on the Care and Management of Laboratory Animals. 5th Edition Churchill Livingstone ch. 28 p. 353.

Specified pathogen free techniques 10

Specified Pathogen Free (SPF) animals may be defined as animals that are free of specified micro-organisms and parasites, but not necessarily free of those not specified. Animals must be drawn from stock that has originally been established by hysterectomy and maintained within a barrier designed to discourage the entry of undesirable organisms. They are cared for by skilled technical staff and maintained on diets, bedding and nesting materials known to be free of pathogenic micro-organisms. Under the Laboratory Animals Centre (LAC) Accreditation Scheme SPF animals are classified as category 3 or category 4.

It should be emphasised at the outset that the SPF concept is one which is achieved by compromise. For example, although all supplies to the unit are sterile the animals are looked after by animal technicians who will inevitably carry micro-organisms, though not necessarily pathogenic (disease causing) ones. The dress and washing procedures of the animal technicians must endeavour to maintain an effective barrier without being inoperable for practical purposes. Here then is the element of compromise or calculated risk.

In the SPF unit the animal technician still has direct contact with the animals, compared with a Germ Free unit where the animals are entirely enclosed within an isolator. The successful operation of an SPF unit depends on a high standard of management to prevent ingress of pathogens.

The results of experiments in which SPF animals are used are likely to be more meaningful due to the lack of interference by overt or sub-clinical disease. There is evidence to indicate that fewer animals need be used in an experiment because of the more reliable results obtained.

Design, operation and construction of the building.

The building is constructed to a high specification to make it virtually air-tight so that only air from the air-conditioning plant is introduced to the animal unit, which is at positive pressure to the atmosphere. All materials entering the unit are sterilised and staff may only enter after taking precautions against carrying in infections. They are usually required to shower thoroughly and change into sterile clothes. Colloquially the inside of the unit behind the shower is known as the 'clean' side, and the outside is the 'dirty' side.

Walls, floors and ceilings are of solid construction with no voids or cavities which could easily be pierced from the inside or outside of the building: this means extensive use of brick and concrete. Windows are usually absent from the animal rooms and are only provided, if at all, in the staff room and then they are double glazed with toughened or even armoured glass. Finishes to walls, floors and ceilings are usually of tough, flexible materials, applied as thick plastic paints or vinyl sheeting, which are easily cleaned and discourage the adherence of dirt. The gaps around doors, windows, pipes, ventilator terminals etc. are sealed with a non-setting mastic which accommodates the slight movement which occurs when disimilar materials such as brick and wood are adjacent.

Rodent barriers consisting of vertical metal sheets with a 5 cm (2 in) lip at the top are placed across the doors at each end of the shower unit to prevent wild or escaped animals entering or leaving the unit. They are usually about 45 cm (18 in) high and can easily be stepped over by staff.

Showers

The shower suite usually consists of five sections. The first forms an air-lock and should contain an insecticide dispenser or electric insect killer to prevent flies entering the unit. The second section is where outside clothing is removed before proceeding to the third section, the shower itself. The spray may be controlled manually or started by closing the door and will run for a pre-set time before the door to the next section can be opened. The fourth section is a drying and dressing area where sterile clothing (supplied daily) is put on before passing to the fifth and final section which is an air-lock leading to the animal area.

The air flow through the shower suite is always from the interior to the exterior and air pressure should be detected when opening the doors on entering the shower suite. When personnel leave the unit they need not shower although they may do for cosmetic reasons. The 'clean' side clothing may be left inside the unit or taken to the 'dirty' side for laundering and re-sterilisation.

Personnel represent the greatest hazard to the SPF unit and anyone with cuts, sores, colds or intestinal upsets, for example, which may result in infectious conditions being transmitted to the animals, should not be allowed to enter. Similarly those who have been in contact with conventional laboratory animals should undergo a 'waiting period' before entering the SPF unit. This period varies from 3 days to 2 weeks depending on local conditions and judgements. Because of the risks staff should be discouraged from keeping pets at home.

Because staff are such a hazard to the unit only those of a high integrity should be selected to work in SPF units. At weekends the unsupervised technician is on trust to carry out the unit's protocol. It is even the practice in some units not to allow the staff who have tendered their resignation to re-enter the unit if there is any suspicion that the barrier protocol will not be observed during their period of notice.

Air filtration

Air entering the unit is sterilised by filtration. The air first passes through a roughing filter which removes gross debris such as insects and large dust particles and requires changing frequently either by pulling a clean section of a roll of filter material into position across the inlet duct or by having two filters in series and changing only one at a time so that all air passes through a roughing filter.

The air passes over a heater battery and usually through a humidifier consisting of a fine water spray or steam jet before entering the fan chamber. Up to this point the air has been sucked through the filters by the fans but having passed through the fan chamber it is then forced through intermediate fine filters and finally through absolute or HEPA (High Efficiency Particulate Air) filters. When the absolute filters are changed it is essential to have two filters in series or parallel so that the inside of the unit never comes into direct contact with partially filtered air.

Air is delivered to the unit at a greater rate than it is removed by the extraction system thus creating a positive pressure. Air should leak out through the shower suite to prevent unfiltered air blowing in. The pressure of the rooms is usually 12 mm (½ in) water gauge to the corridors which are at 12 mm (½ in) water gauge to the outside. Air is extracted from the animal room through a medium filter which retains dust and fur, by a fan in the exhaust duct which blows the air through a non-return flap or damper which closes automatically in the event of a failure of the air supply to prevent unfiltered air entering the unit.

Water treatment

Some users consider that as ordinary chlorinated tap water is fit for human consumption it is adequate for SPF animals. Others prefer to add more chlorine to a level of 15 ppm or to acidify the water to pH 2.5. More sophisticated water treatments involve sterilisation by filtration through porous ceramics or by using ultra-violet (UV) light.

Piped 'on tap' disinfectant systems for washing down rooms and floors are of dubious value; apart from the very high cost of adding the disinfectant there is evidence to suggest that certain micro-organisms thrive even in high concentrations of the disinfectant solution.

Diet sterilisation

There are three effective methods of sterilising diets:
Autoclaving
Fumigation
Irradiation

Autoclaving

The autoclave may be programmed to provide a high temperature, short exposure cycle (130°C for 1–2 min) or a lower temperature, long exposure cycle (121°C for 10–15 min). The first method is preferable as the total amount of heat applied to the diet is less and the damage to nutrients is less in consequence. The procedure for autoclaving diet is to spread the pellets on shallow wire trays to a depth of 2.5–5.0 cm (1–2 in) to allow rapid steam penetration. If the autoclave is a high vacuum machine it is beneficial to draw a vacuum for 10 minutes after the sterilising cycle in order to dry the diet. Autoclaving may damage nutrients by destroying vitamins and denaturing proteins and pellets may be physically damaged by the time sterilisation temperature is reached throughout the pellet.

Fumigation

Fumigation employs a modified autoclave into which a sterilising gas such as ethylene oxide is introduced. A serious disadvantage is that the diet has to be ventilated for up to 3 days before it is free of fumigant and there is a suspicion that some toxic compounds are formed. Therefore this method is not widely used.

Irradiation

The most common method of sterilisation is irradiation using gamma rays produced by the radioactive isotope ^{60}cobalt (^{60}Co). These highly penetrating rays are effective for destroying micro-organisms without causing the exposed materials to become radioactive. The unit of measurement of radiation is the rad and diets intended for SPF animals usually receive a dose of 2.5 megarads (Mrads). Diets for germ free animals are generally passed through the irradiation plant twice and so receive a total dose of 5.0 Mrads.

An important advantage of this method is that the nutritional and physical damage to the diet is minimised as the temperature rise is limited to 2 or 3°C. The diet is packed in a plastic bag (12.5 kg is usual) which is evacuated to prevent oxygen reacting with the gamma rays and then sealed. This bag is then sealed in a second bag and finally placed in a fibre-board box or paper sack. The package is now passed between the radioactive sources within the irradiation plant. A previously applied yellow or orange indicator sticker should have turned dark red when the package emerges indicating that it has received a dose of 2.5 Mrads. (Diets intended for germ free animals receive a second sticker and the process is repeated, the doses being cumulative). The high capital cost of the irradiation plant and the expenses of double wrapping, packaging and transportation are reflected in the market price.

On receipt at the SPF unit the packages should be carefully handled and stored to prevent any of the wrappings becoming punctured or damaged. The diet is passed into the unit as follows:

There should be two operators on the outside ('dirty side') of the unit wearing sterilised outer clothing and gloves. At a point very close to the dunk tank the package is opened by an operator who must only handle the outside of the package and who checks the indicator sticker is dark red. The second operator withdraws the double wrapped bag of diet and checks that the inner bag feels hard and is therefore still under vacuum. The outer bag is opened and the inner bag containing the diet carefully slipped into the dunk tank. The bag of diet is passed through the germicidal solution and is withdrawn by staff inside the SPF unit. Care should be taken to prevent trapped air bubbles reaching the inside of the SPF unit. Diet which appears wet indicates a punctured wrapping and must not be taken into the SPF unit.

Drains and lavatories

Water may be used in SPF units for hosing down and general cleaning purposes. A standard drainage system is used with an extra water trap on the main exit pipe. If a drain remains unused for an extended period the water may evaporate sufficiently to break the seal and allow organisms from the sewer to enter the unit (though the second trap should prevent this). Water poured down the unused drain once a week will maintain the seal. Alternatively a small quantity of light oil poured onto the surface of the water will prevent evaporation. Dry cleaning systems where few drains are used considerably reduce this problem.

It is generally accepted that providing lavatories inside the SPF unit presents no great risk or contamination if staff observe high standards of hygiene and is almost certainly less than if they had to leave the unit and return. The flushing mechanism is preferably operated by a foot button.

Waste materials

Apart from waste water, which has already been considered, the waste materials from an SPF unit are dead animals, general refuse and large quantities of soiled bedding. These are usually passed out through the autoclave but not sterilised. When the autoclave has been unloaded from the clean side following a sterilisation cycle, the outgoing refuse may be loaded in and passed out. In some instances the exit port used for issuing animals may be used. Refuse from an SPF unit is, of course, as microbiologically 'clean' as the unit itself.

Issue of animals

Boxes for dispatching animals from the SPF unit are usually cardboard or plastic with a pad of fine filter material stapled or taped over the wire mesh vents. The whole box is autoclaved into the unit, animals for issue placed inside, the lid sealed on with tape and the whole box passed out through a special exit lock fitted with an ultra-violet lamp. Alternatively a single flap door may be used which relies on the outrush of air to maintain the barrier when the box is passed out.

Sterilisation of new SPF buildings

After the building has been handed over by the contractor it should be thoroughly cleaned and washed down. The ventilation system should be run at full power before the absolute filters are fitted to blow any dust out of the ducts. The absolute filters should then be installed. Walls, floors, ceilings and the outside of ventilation ducts should be thoroughly swept down. This should be followed by a further wash down with a hot solution of 4% washing soda which will substantially reduce the microbiological load and remove any oil or grease that may be present.

The dunk tank should be filled with the liquid sterilant to be used. All doors except those to the outside air-lock of the shower suite should be propped slightly open to allow the fumigant to penetrate the shower unit. The external intake and exhaust grills of the air conditioning system should be sealed with polythene sheet attached with PVC tape. (All the trunking of the air conditioning system should be gas tight but this should be checked with the site engineer.) A sterilised set of clothing should be placed in the inner changing room ready for the first entrant after fumigation.

The total volume of the SPF unit should be calculated to determine the amount of formalin solution (37% solution of formaldehyde gas in water and stabilised with methyl alcohol) that will be required to be evaporated. (Useful rule of thumb figures are 1 ml of a 10% formalin solution per cubic foot, or 35 ml per cubic metre.)

The safest and most convenient method of evaporating formalin is to use electrically heated vessels, fitted with thermal cut-out devices, strategically placed around the unit. Prills which release formaldehyde when heated may also be used and require their own appropriate type of heating vessels (see chapter on sterilisation and disinfection). A simple method of switching on all these vessels at once is to use the main switch for the power circuits (not lighting) situated outside the SPF area.

The heating vessels should be filled with the appropriate quantity of formalin or prills and if possible the relative humidity in the unit raised to about 80%. (The most practical way of achieving this is to wet the floors with hot water and allow to evaporate.) The exit doors to the unit should be closed and sealed with tape and warning notices displayed. The main switch is turned to 'on' and fumigation proceeds. After a few minutes fumes should be visible through the viewing window. The building should be left for at least 24 hours before it is purged of fumigant. This is done by removing the sealing cap from the intake to the air conditioning system and switching on the intake fan. After a few minutes the unit becomes pressurised and the sealing cap is removed from the exhaust grill: finally the exhaust fan is turned on. One or two days should elapse before entering the unit.

A normal SPF entry can be made taking care that the outer door to the first air lock is shut before the second door is opened leading to the changing room. The usual showering routine is followed, unpropping and closing each door as the entry proceeds. The previously supplied sterile clothing is put on in the inner changing room. All equipment, such as racking, cages and domestic supplies, is passed into the unit via the autoclave or dunk tank as appropriate. The unit is now ready for use.

(NB. If the unit forms part of a larger building considerable caution should be exercised and investigation undertaken *before* fumigation as there is a risk of fumes leaking to other parts of the building.)

Introduction of animals

Animal colonies are established within the unit by introducing stock from an SPF or germ free source. Further pathogen-free young obtained by hysterectomy of pregnant animals outside the unit are passed aseptically through the barrier and are fostered by the previously introduced pathogen-free females.

Microbiological screening

A sample number of animals are submitted to a pathology laboratory for screening. Alternatively swabs from animal faeces, walls or floors or samples of animal faeces or water supply may be sent for culturing. A Petri-dish of culture medium may be left open in the unit to monitor the air for organisms before being resealed and again sent for culturing. Submitting an animal for microbiological testing is the most useful and logical method as this is the product of the SFP unit.

If on-site facilities for culturing are not available the Medical Research Council Laboratory Animals Centre, regional Veterinary

Investigation Centres, Local Public Health Laboratories or commercial diagnostic laboratories will provide this service.

Clothing

A variety of suitable clothing is available. Underwear with shirt, trousers or a boiler suit are common, together with a paper hat which covers the hair. Face mask and surgical gloves may also be worn. Footwear usually consists of lightweight wellington boots or plimsolls.

The mouse (*Mus musculus*) 11

Body temperature	37.4°C (99.5°F)
Age at sexual maturity	5 weeks
Age at first mating	6 weeks
Oestrous cycle	4–5 days
Duration of oestrus	10–20 hours
Gestation	19–21 days
Litter size	8–11
Weaning age	18–21 days

The mouse is the most commonly used experimental animal, probably because of its small size and fecundity. A mouse can produce a litter every 3 weeks and 5 generations can be achieved in one year. There are many inbred and outbred strains in a variety of coat colours; these include leukaemic, spontaneous tumour bearing, obese, blind, tail-less and athymic nude. The majority of mice are used for cancer research.

Anatomy

The mouse is a rodent exibiting the typical dentition of this order (Fig 12.1). Only the female has nipples, a fact which can assist sexing the very young animal prior to their becoming obscured by hair growth.

Reproduction

Oestrus last for 10–20 hours in the 4–5 day oestrous cycle. After mating, a copulation plug may be seen in the vagina or on the droppings tray of a wire floor cage. Gestation is 19–21 days and the young are born blind and naked. Fur is seen at 5 days of age, the eyes open at 14–15 days, and the ears open at 14 days. The young may be weaned at 18–21 days. Mice come back into breeding conditions immediately after giving birth to their young and if kept with a male mating is likely to occur at the post partum oestrus. It is then essential to wean the first litter before the second is born or the newborn litter may be smothered, or cannibalism may occur. Pseudopregnancy lasts for 12–14 days and may be induced by a sterile mating.

11.1. *An obese mouse*

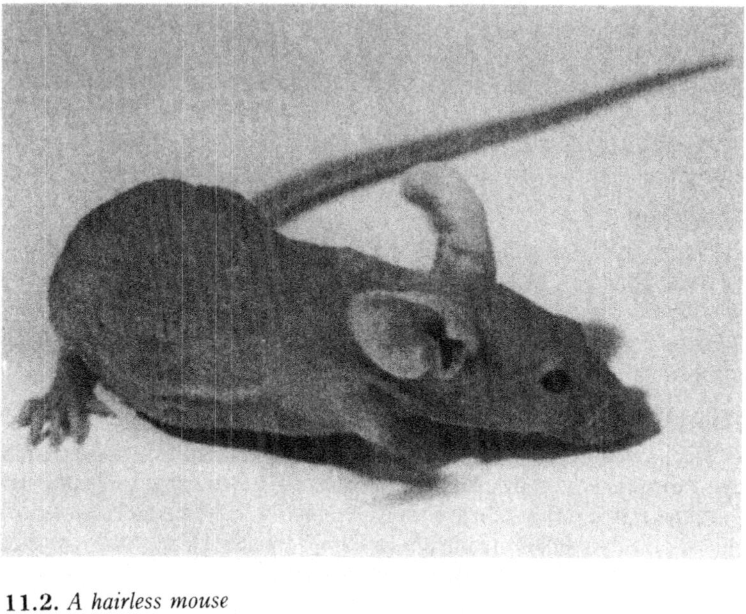

11.2. *A hairless mouse*

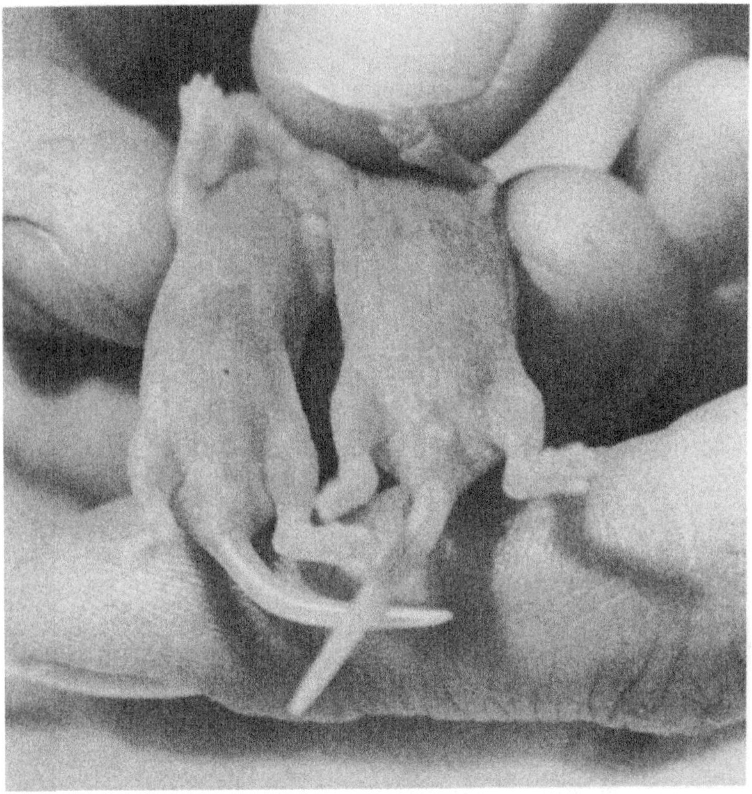

11.3. *Sexing day old mice, female left, male right*

Nutrition

Food and water should be available at all times. A diet containing 18–20% protein for breeders and 15–16% for stock is given in a pellet about 25 × 10 mm diameter (1 × 3/8 in). Some black strains have difficulty gnawing hard pellets, possibly due to congenital tooth malformation.

Handling

The mouse should be picked up by the scruff of the neck using thumb and forefinger, or by holding the base of the tail with one hand and gripping the scruff with the other as described. With the mouse held this way, the handler is unlikely to be bitten and the animal is secure. A refinement of this technique allows one hand free for dosing or injecting.

Identification

White or light coloured mice may be individually identified by using fabric or histological dyes as a code on the coat. It has been found that some green dyes cause hair loss and picric acid is unsuitable as it is believed to be carcinogenic. Mice may be identified by a tattoo at the base of the tail using a specialised instrument, or by punching a code of holes in the ears with a chicken toe punch (Fig 12.7).

Environment

The room temperature should be about 21°C (70°F) with a relative humidity of about 50%. There should be a definite lighting regime of approximately 12 hours light and 12 hours darkness for breeding animals. Permanent light gives rise to anoestrus and the females will not allow mating. Natural light through windows may cause seasonal variation in breeding performance. Some strains of mice do not breed well when kept in close proximity to another particular strain, the reason for this is not known but pheromones may be involved.

Caging and bedding

Breeding pairs or trios are usually housed in cages 25 × 13 × 13 cm (10 × 5 × 5 in) harems (1 male – 5 females) or stock mice (15–20) in

11.4. *Sexing adult mice, female left, male right*

11.5. *Restraining a mouse*

Common diseases

Type	Common name	Causal agent
Bacterial	Salmonellosis	*S.enteritidis, S.typhimurium*
	Tyzzer's disease	*Bacillus piliformis*
Viral	Ectromelia – mousepox	virus
	Enteric mouse hepatitis virus (EMHV)	virus
	Mouse hepatitis virus (MHV)	virus
Endoparasitic nematode	Round worms	*Aspicularis tetraptera* and *Syphacia obvelata*
Ectoparasitic mite	Body mange	*Myocoptes musculinus*
	Head and neck mange	*Myobia musculi*

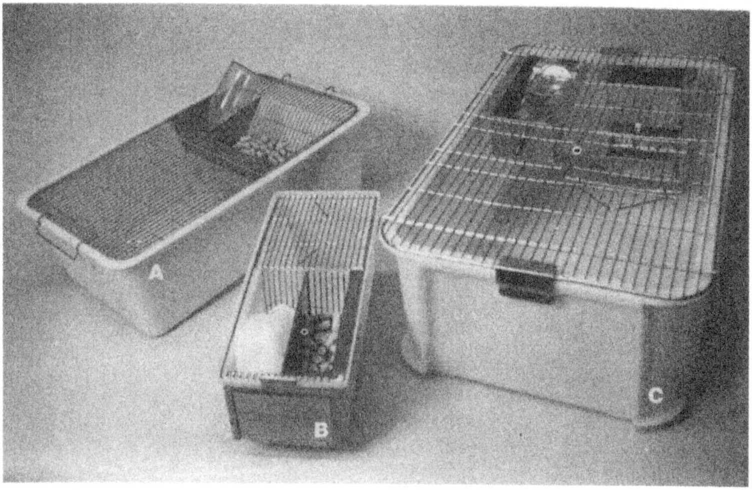

11.6. *Polypropylene rodent cages (a) rat (b) mouse (c) guinea -pig*

cages 50 × 25 × 15 cm (20 × 10 × 6 in), contructed of polypropylene or other heat stable plastics with a wire mesh top designed to hold food and a water bottle (Fig 11.6). Soft whitewood shavings or sawdust may be used as bedding material and shredded paper for nesting if necessary.

The Rat
(Rattus norvegicus) 12

Body temperature	37.5°C (99.5°F)
Age at sexual maturity	40–50 days
Age at first mating	70 days
Oestrous cycle	4–5 days
Duration of oestrus	10–20 hours
Gestation	21–22 days
Litter size	9–11
Weaning age	18–21 days

The rat is an intelligent animal easily tamed by handling. There are inbred strains such as PVG/C, WAG, Lewis and DA. The non-inbred strains include Wistar, Sprague-Dawley and hooded Long Evans. They are used extensively for nutritional and toxicological studies.

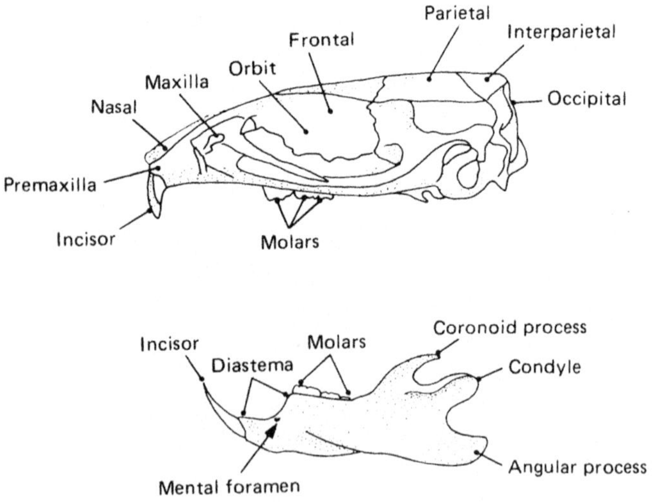

12.1. *Rat skull*

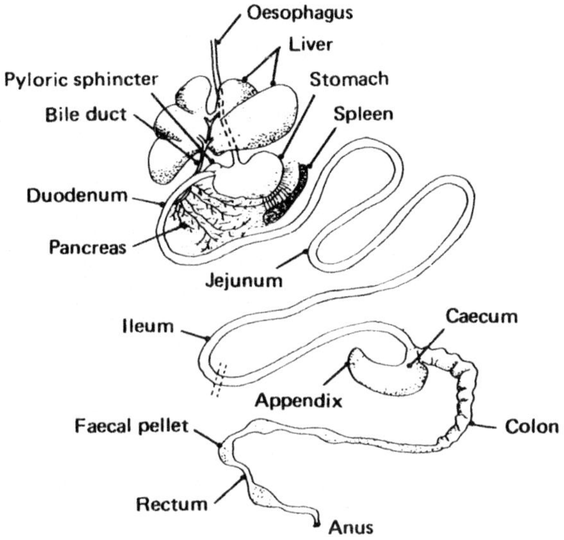

12.2. *Rat alimentary canal*

Anatomy

The rat is a rodent with typical dentition for this order (Fig 12.1). Like the mouse only the female has nipples which are a guide to sexing very young animals. The rat does not have a gall bladder, bile is produced in the liver and passes directly to the duodenum via the bile duct.

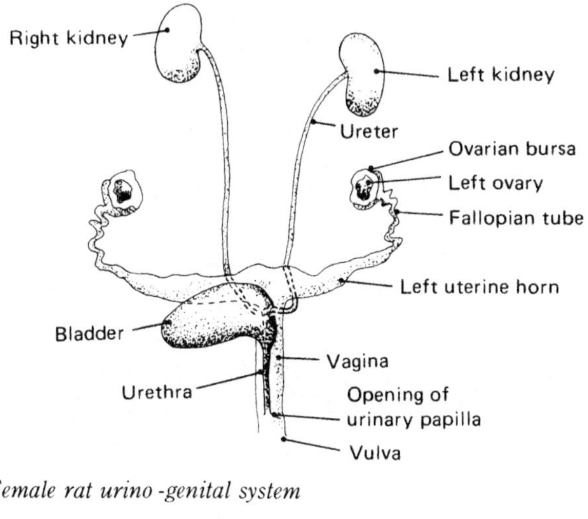

12.3. *Female rat urino -genital system*

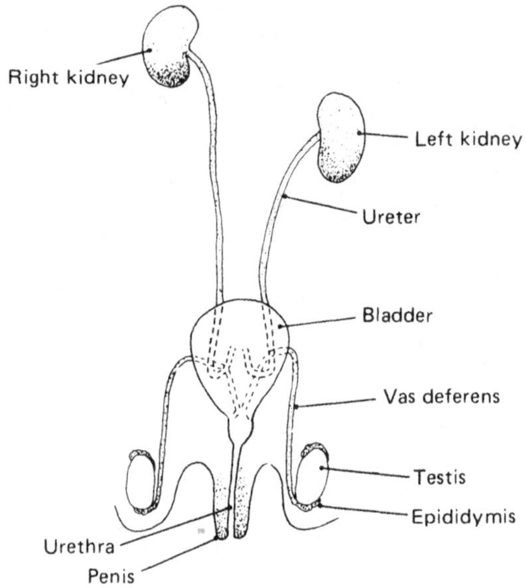

12.4. *Male rat urino -genital system*

Reproduction

Oestrus lasts for 10–20 hours of the 4–5 day oestrous cycle. A straw coloured copulation plug may be seen in the vagina or because of its size on the droppings tray of a wire bottomed cage. Gestation is 21–22 days and the young are born blind and naked. Rats may be kept in monogamous pairs all their breeding lives thereby taking advantage of the post-partum oestrus, or in harems with one male to 6–8 females, removing the females to separate cages when they are obviously pregnant and returning them to the harem after the young are weaned.

Nutrition

Food and water should always be available. Breeding animals require a diet containing 18–20% protein and stock 15–16% given in pellet form approximately 25 × 10 mm (1 × 3/8 in) in size. Many commercially available rat diets are suitable for breeding and maintenance.

Handling

The rat should be picked up by placing one hand over its back and grasping it firmly round the neck so that it cannot turn its head to bite. When animals become tamed they can be picked up by holding them around the shoulders. A pregnant animal should be given additional support by cupping the other hand round the lower part of the animal's body and legs.

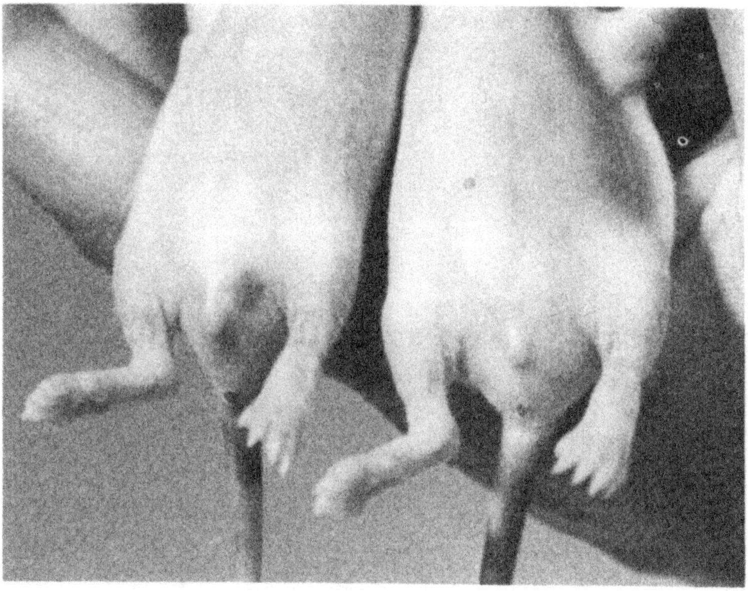

12.5. *Sexing 10 -day -old rats, male left, female right*

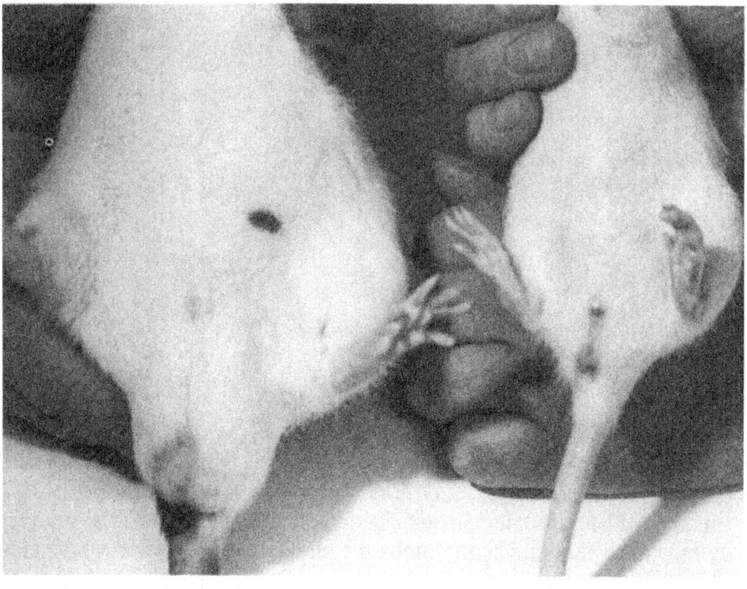

12.6. *Sexing adult rats, male left, female right*

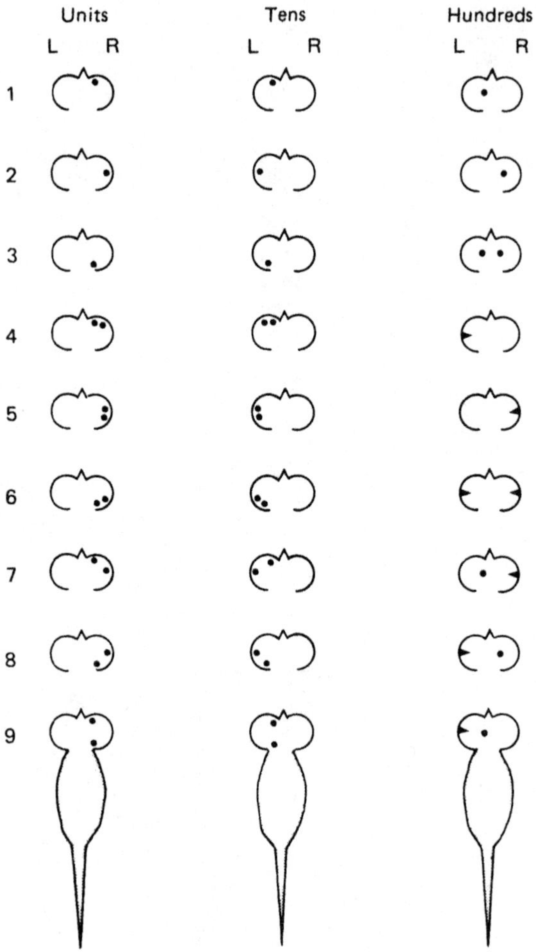

12.7. *Punched hole ear code for rat or mouse*

Identification

The ears are large enough for a punched hole code to be formed using a chicken toe punch (Fig 12.7). Rats may also be tattooed in the ear or at the base of the tail using a specialised instrument. Fabric or histological dyes can be used for light coloured animals but green dyes may cause hair loss and picric acid is believed to be carcinogenic. Numbered ear tags can be inserted with special pliers but tend to be torn out.

Environment

The room temperature should be about 21°C and the relative humidity at about 55%. Low humidity is thought to cause 'ringtail', a condition where the rat's tail becomes ringed or constricted and in extreme cases can slough off. In a very dry atmosphere the humidity may be increased by leaving floors wet after washing or by placing open pans of water in the room.

Caging and Bedding

Breeding pairs can be housed in boxes with a minimum height of 25 cm and an area of 55 × 27 cm, with a wire mesh top to hold food and water bottle (Fig 11.6). Harems and stock (e.g. 6 adults) should be housed in larger boxes, harems and animals over 100 gms need 250 cm^2 per animal and stock up to 100 g 150 cm^2 per animal. Solid floors or wire grids with a tray for faeces below are both suitable the latter allowing greater air flow through the cage thereby decreasing the immediate humidity. White softwood shavings or sawdust can be used for bedding material and shredded paper or woodwool for nesting.

Common diseases

Type	Common name	Causal agent
Bacterial	Labyrinthitis (Middle ear disease)	*Streptobacillus moniliformis*
	Paratyphoid	*Salmonella enteritidis* *Salmonella typhimurium*
Viral	Chronic respiratory disease (CRD)	
	Endemic pneumonia	A mycoplasma and a virus
Endoparasitic nematode	Round worm	*Syphacia obvelota*
Ectoparasitic mite	Mange	*Myobia ratti* *Myobia musculi* *Notoedres spp*
Physiological	Ringtail	Possibly low humidity

The Guinea-Pig *(Cavia porcellus)* 13

Body temperature	38.3–40°C (100–104°F)
Age at sexual maturity	60–80 days
Mating weight	500 g
Oestrous cycle	14–16 days
Duration of oestrus	6–11 hours
Gestation	59–72 days
Litter size	3–6
Weaning age	14–21 days

Guinea-pigs have been extensively bred by pet fanciers and there are now three main varieties; English (smooth hair), Abyssinian (rosetted hair) and Peruvian (very long hair). There are many colour variations of these and new ones are constantly being bred. The most common laboratory strain in this country is the white or cream Dunkin-Hartley, although there are now large populations of pink-eyed white (albino) animals.

Anatomy

The guinea-pig is an herbivorous rodent, a fact which is reflected in its anatomy. It has rodent dentition and the alimentary canal is typical of a herbivore, having a large caecum for cellulose digestion (Fig 13.1). There are two mammary glands and the nipples, situated in the groin, are found in both sexes. The guinea-pig has no tail, has 4 toes on the front foot and 3 on the hind foot.

Reproduction

There is no defined breeding season but in an environment influenced by natural light seasonal variation may be apparent. Guinea-pigs are commonly bred in harem systems of 1 male to 5–7 females. Oestrus can be observed by the degree of vaginal opening and copulatory plugs, usually found on the floor of the cage, will indicate mating has taken place. By recording these signs mating can be timed. When the guinea-pig is not in oestrus the vagina is normally sealed by a

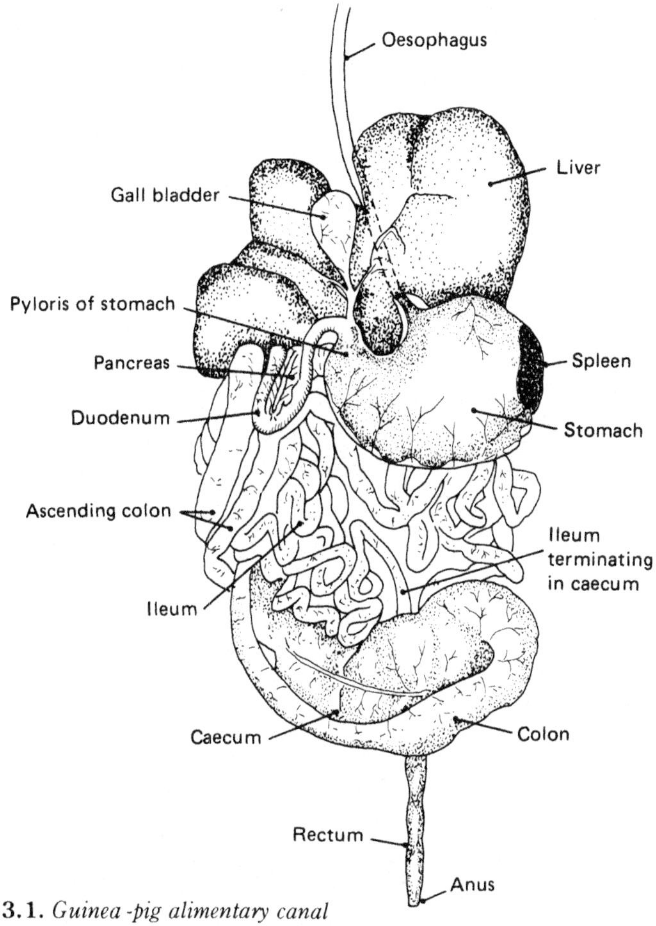

13.1. *Guinea -pig alimentary canal*

membrane. However it is possible to rupture this and it is therefore not a foolproof indication of the stage of oestrus.

Guinea-pigs exhibit a post-partum oestrus and if the harem is left intact further mating will occur. The young are well developed at birth and will suck any lactating female in the harem. Accurate breeding records are difficult to keep with this system. Young are normally weaned at 14 days but with care can be weaned from birth.

Nutrition

The most important feature of guinea-pig nutrition is that like primates it is one of the few species unable to synthesise its own vitamin C; this must be provided in the diet or drinking water as ascorbic acid.

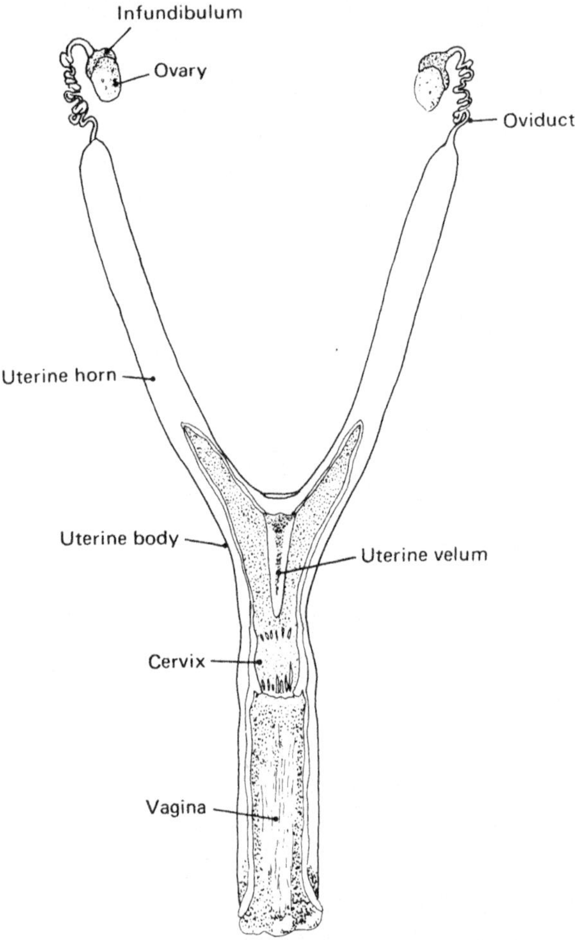

13.2. *Female guinea-pig reproductive tract (partly dissected)*

Vitamin C is destroyed by contact with metal and therefore only glass drinking spouts should be used. Fresh green food can be used to provide vitamin C but may often be contaminated with pathogens; 50 g of cabbage will provide a guinea-pig's daily need. Guinea-pigs eat about 25–50 g of pelleted diet per day. They drink approximately 85 ml of water per day but are messy and wasteful with water and bottles are quickly fouled.

Handling

Because they so rarely bite guinea-pigs tend to be mishandled. When they are disturbed they will race around the cage. A guinea-pig should

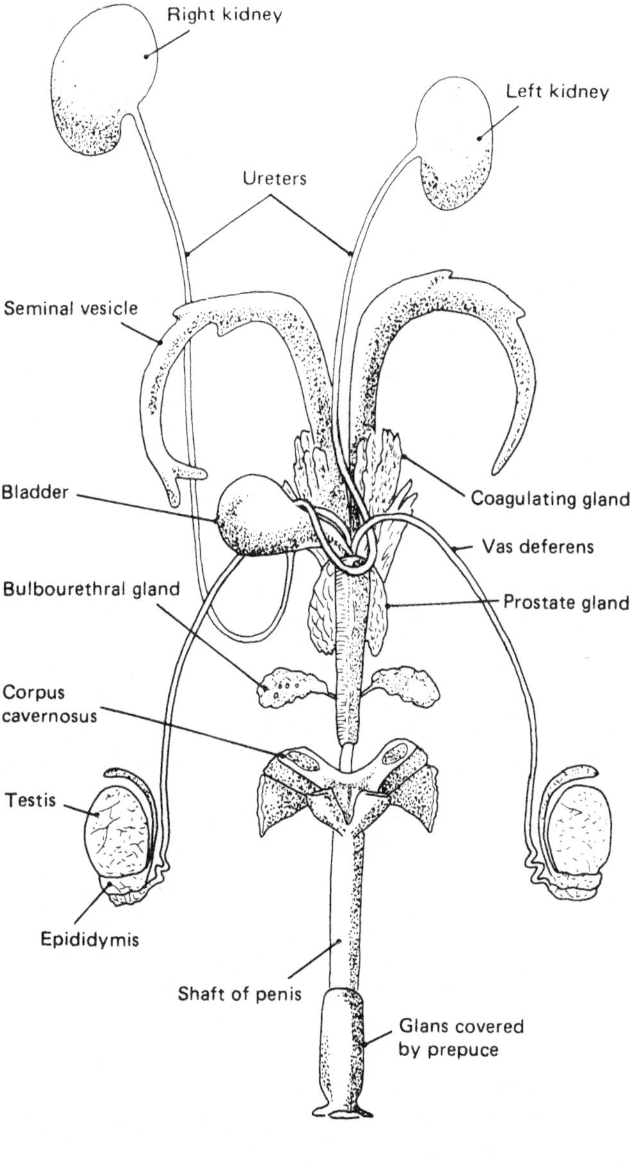

13.3. *Male guinea -pig urino -genital system*

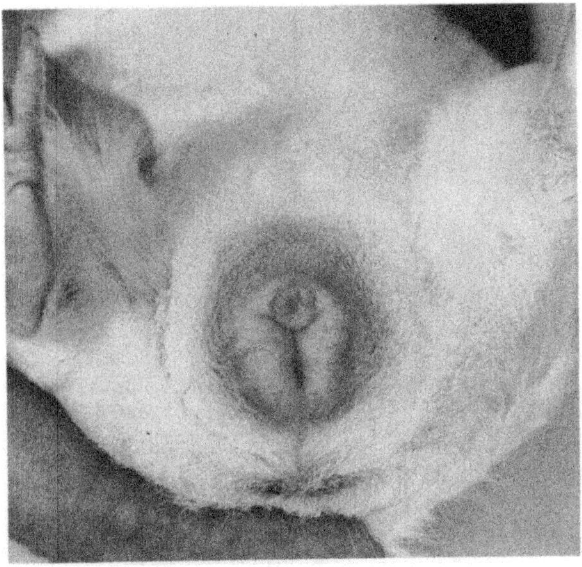

13.4. *Sexing adult male guinea -pig*

be caught by placing one hand in front of the animal and one behind; it can then be lifted from the cage with the body supported in both hands.

Identification

Identification can be a problem: ears are often coloured making tattooing difficult, tags tend to be torn out and notches and holes in the ear soon become confused with tears. In albino colonies ear tattooing is probably best (Fig 15.8); however for coloured animals a punched hole is the only practicable method.

Environment

Guinea-pigs require a room temperature of about 18°C (65°F). Breeding performance, for example, may be affected by high summer temperature. Windowless rooms with time controlled artificial lighting are advisable to prevent seasonal effects on breeding.

Caging and bedding

The most common system is the wire floored cage with manual or automatic cleaning and/or flushing underneath. If solid floors are used softwood shavings and hay are suitable bedding and nesting materials (Fig 11.6). Automatic drinkers are widely used; these should be tamper proof and strong (preferably stainless steel) as guinea-pigs will constantly chew and play with them (Fig 1.2).

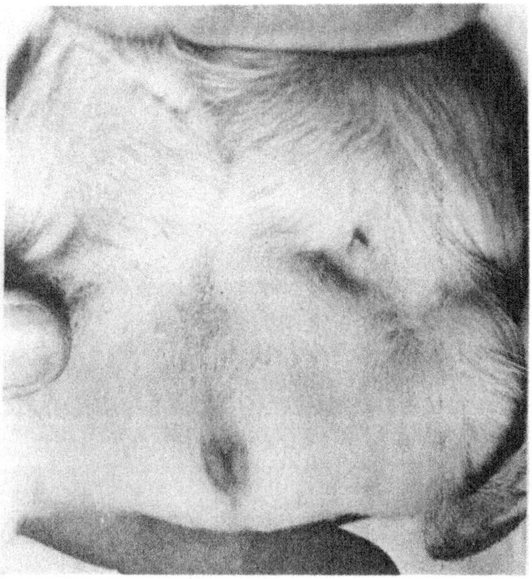

13.5. *Sexing adult female guinea -pig*

Disease

It is important to note that most antibiotics are reported to be toxic to guinea-pigs (especially penicillin, streptomycin and terramycin) and should not normally be used for treatment of this species.

Common important diseases are:

Type	Common name	Causal agent
Bacterial	Salmonellosis	*Salmonella typhimurium*
	Pseudotuberculosis	*Yersinia pseudotuberculosis*
	Streptococcal pneumonia	*Streptococcus pyogenes*
	Streptococcal lymphadenitis	*Streptococcus pyogenes*
Endoparasitic protozoal	Coccidiosis	*Eimeria caviae*
Ectoparasitic lice		*Gyropus ovalis* *Gliricola porcelli* *Trimenopon jenningsi*
Physiological	Intrauterine haemorrhage	obscure dietary or liver metabolism imbalance
Dietary		Lack of vitamin C

The Syrian hamster (*Mesocricetus auratus*) 14

Body temperature	37.4°C (99.3°F)
Age at sexual maturity	6 weeks
Mating weight	100 g
Oestrous cycle	4 days
Duration of oestrus	12 hours
Gestation	16 days
Litter size	7–9
Weaning age	21 days

14.1. *Syrian hamster*

14.2. *Chinese hamster (Shell Research, Sittingbourne)*

The Syrian hamster, *Mesocricetus auratus*, is also known as the 'golden' hamster although this term is rather inaccurate because coat colours within this strain vary considerably. The Syrian hamster as its name implies reached this country from Syria where three pairs were captured in 1937 in Aleppo (Fig 14.1). Hamsters are used mainly for cancer research, virological and toxicological studies.

Anatomy

An adult Syrian hamster weighs about 100 grams and coat colours include albino, golden, ginger and cream. Hamsters possess cheek pouches in which they gather food for later storage in their nests.

The Syrian hamster should not be confused with the large European hamster or the Chinese hamster which is smaller, grey in colour with a dark stripe running down its back and has a short but almost prehensile tail (Fig 14.2).

Reproduction

The hamster is a seasonal breeder and comes into oestrus from February to October; in a windowless room with controlled lighting giving at least 12 hours light it is possible to reduce seasonal variation. Hamsters are most active in the dark and because of this the lighting cycle can be reversed to provide darkness during the working day to facilitate the special mating system. The female is placed with the male and the pair are observed to check if mating takes place. Mating will take place immediately if the female is in oestrus, but if the female does not readily accept the male she is not in oestrus and should be removed before fighting occurs; this needs care by the handler because of the danger of being bitten. This method of mating can be time consuming and harems or monogamous pairs may be used in preference. If hamsters are grouped at weaning fighting rarely occurs.

On average a ratio of 1♂ to 5♀ is used in the harem and the females should be removed to a littering cage a few days before parturition and returned to the harem after the litter is weaned.

Young hamsters are approximately the same size as young mice but they have a shorter tail. At about 12–14 days the eyes open and the young become active and grow steadily towards independence and weaning at 21 days.

Nutrition

Laboratory hamsters are usually fed on compounded diets made into pellets; a typical rat and mouse diet is suitable. Supplements of sunflower seeds and a variety of cereals are also enjoyed by hamsters but it is not advisable to feed fresh vegetables because disease such as pseudotuberculosis may be introduced to the colony. Water should always be available from bottles or automatic drinkers in preference to open dishes which easily become soiled.

Handling

A frightened hamster often appears aggressive which it demonstrates by rolling on its back or by standing on its hind limbs and screaming at the handler. This is little more than a defence mechanism and should be ignored. The animal may be lifted in cupped hands if it is used to being handled or by firmly grasping the loose skin at the back of the neck. The latter method of handling is preferred if manipulation is necessary; the hamster should be supported so that the fingers may be wrapped around its legs ready for examination, injection or dosing.

Identification

There are not many successful methods of identifying hamsters. Coat colour can often serve as a method but becomes confusing when dealing with large numbers. Tattooing the ear with forceps is a difficult and time consuming job. Ear punching may be considered a useful method but the marks can be obscured by fighting.

Environment

The optimum temperature is 20–21°C (68–70°F) with a relative humidity of 50–60%. Hamsters need at least 12 hours of light, which may be artificial, if they are to be bred successfully throughout the year.

Caging and bedding

Rat or mouse cages are equally suitable for hamsters. The addition of an activity wheel will provide hours of exercise and helps to prevent the animals becoming bored and destructive. Breeding cages should contain a small nesting area set aside to encourage good mothering although this is not a prerequisite. Absorbent materials such as sawdust or woodshavings should be used on solid floor cages and may serve as both bedding and nesting material. Pregnant females will spend a considerable amount of time building nests out of woodwool or shredded paper.

Common Diseases

Type	Common name	Causal agent
Bacterial	Salmonellosis	*Salmonella enteritidis* and *Salmonella typhimurium*
	'Wet Tail'	*Escherichia coli* and *Proteus vulgaris*
Ectoparasitic mite	Body mange	*Demodex criceti*

The rabbit (*Oryctolagus cuniculus*) 15

Body temperature	38.3–39.5°C (100–103°F)
Age at sexual maturity	4–5 months depending on size
Mating weight	variable according to breed
Oestrous cycle	continuous
Duration of oestrus	–
Gestation	30–32 days
Litter size	6–8
Weaning age	4–6 weeks

The rabbit has been bred for many years for meat and by pet fanciers. A large number of breeds now exist and several are used for experimental work. The breeds can be grouped according to body size: small, medium and large. Breeds from each group can be found in the laboratory – small, Polish 1–1½ kg, Netherland Dwarf 1½ kg; medium, Dutch 1½–2½ kg; large, Flemish Giant, 6 kg, New Zealand White 5 kg, Californian, 4½ kg.

Anatomy

Rabbits are not closely related to rodents and their resemblance is only superficial. In both groups the incisor teeth grow continuously; rabbits however have a smaller second pair. They have three premolars in the upper jaw and two in the lower (Fig 15.2). The molar teeth have cutting edges whereas in rodents they are modified for grinding. Both the male and female have nipples, normally four pairs. The alimentary tract has a large caecum and a well developed appendix, as might be expected in an animal eating a high proportion of cellulose. The pancreatic duct enters the duodenum well below the bile duct (Fig 15.3).

Reproduction

Mating induces ovulation in rabbits and unless mated they will remain in oestrus indefinitely. Two principal breeding methods are used,

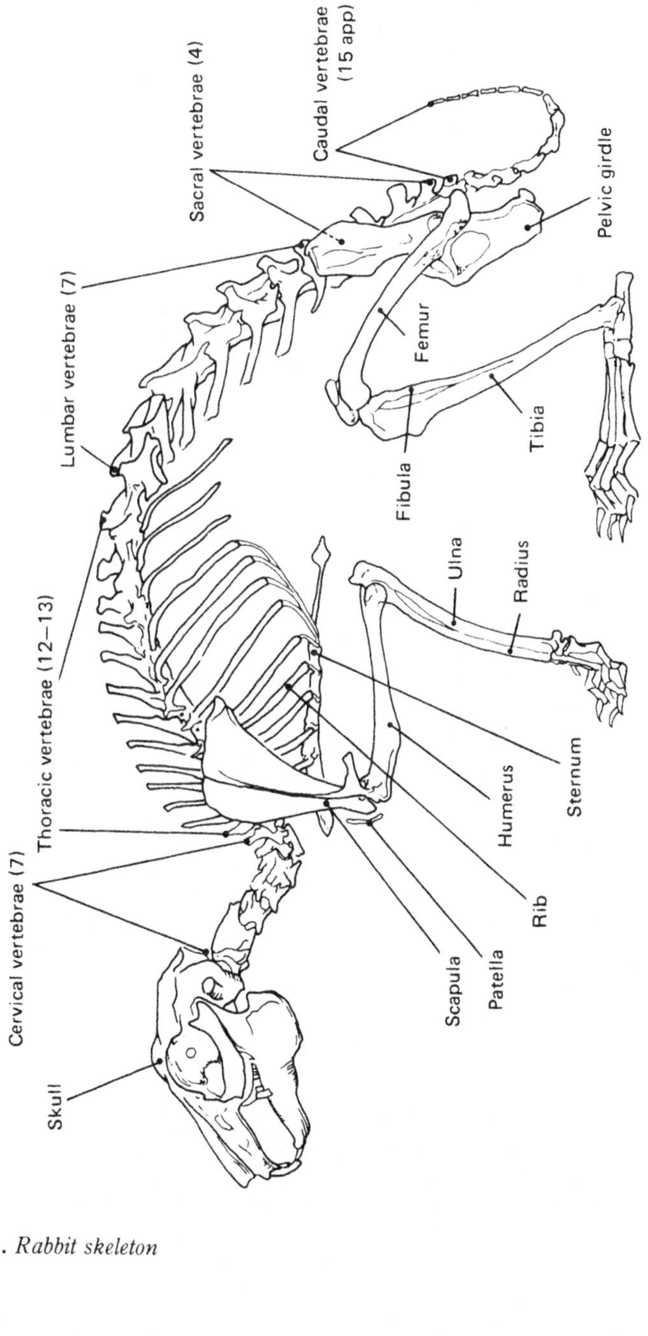

15.1. *Rabbit skeleton*

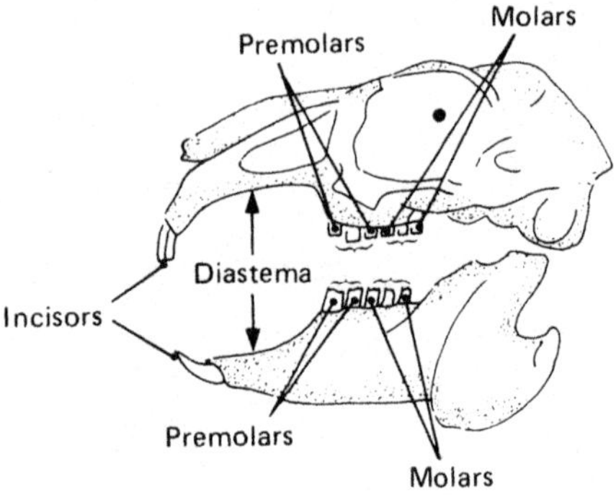

15.2. *Rabbit skull*

females are mated either during lactation or after the young are weaned. Contrary to popular belief rabbits are not easy to breed; the doe's mating, breeding and rearing performance is influenced by a wide range of environmental stimuli.

Females are usually first mated at five months of age and subsequently during the third week of each lactation or after the young are weaned. The female should be taken to the male for mating and they are normally left together until copulation is observed; a second mating a few hours later is sometimes recommended.

Young from females mated at the third week of lactation should be weaned at six weeks to give an interval of at least one week before the next litter is born. Nest boxes should be available 3 or 4 days before

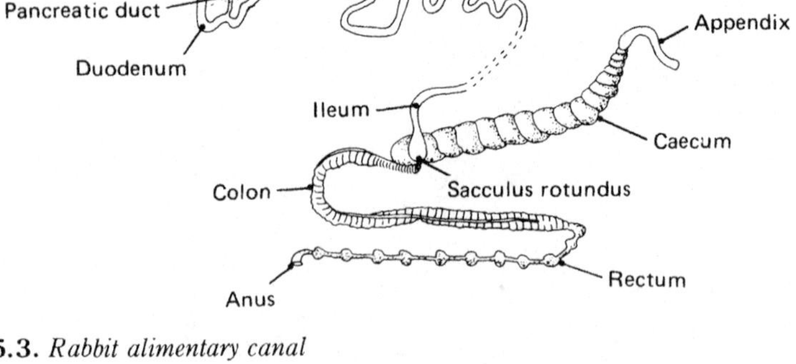

15.3. *Rabbit alimentary canal*

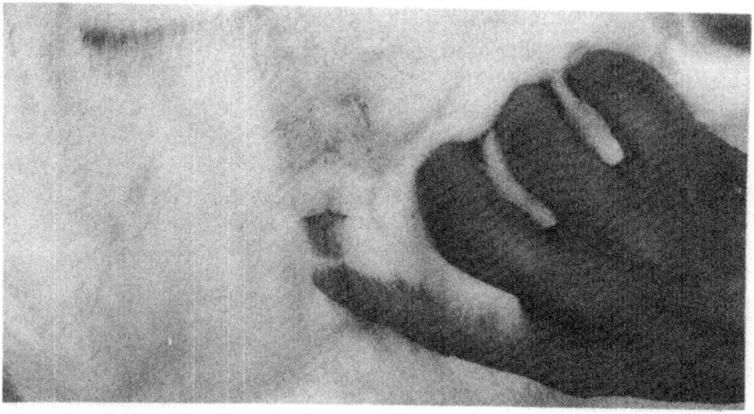

15.4. *Sexing adult female rabbit*

parturition. It has been found that poor mothering tendencies during the first few days after parturition can sometimes be overcome by removing the nest box and young until the next morning. Rabbits only feed their young once in every twenty-four hours, usually early in the morning.

Nutrition

Food and water should always be available. Rabbit feeding practice has evolved largely from the experience of those keeping rabbits for

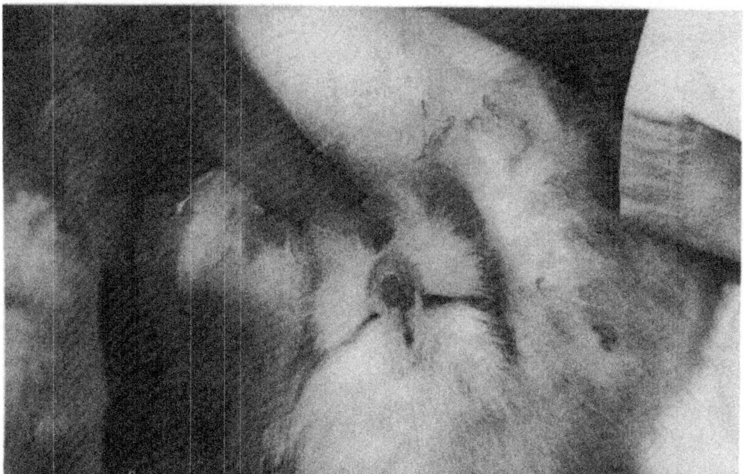

15.5. *Sexing adult male rabbit*

meat and fur. The feeding regime will depend on whether the rabbits are growing, breeding or lactating. For maintenance, diets containing 12–15% digestible protein, 20–70% crude fibre, 2–3.5% fat and 43–47% carbohydrate have been recommended. Pregnant animals need an increase in protein and energy and a reduction in fibre. During the fourth week of lactation two to three times the energy allowance for maintenance is recommended. Examples of pelleted diets available for rabbits are SG1, Diet 18, RAG and R14. These diets are similar to those for guinea-pigs, but are not fortified with Vitamin C.

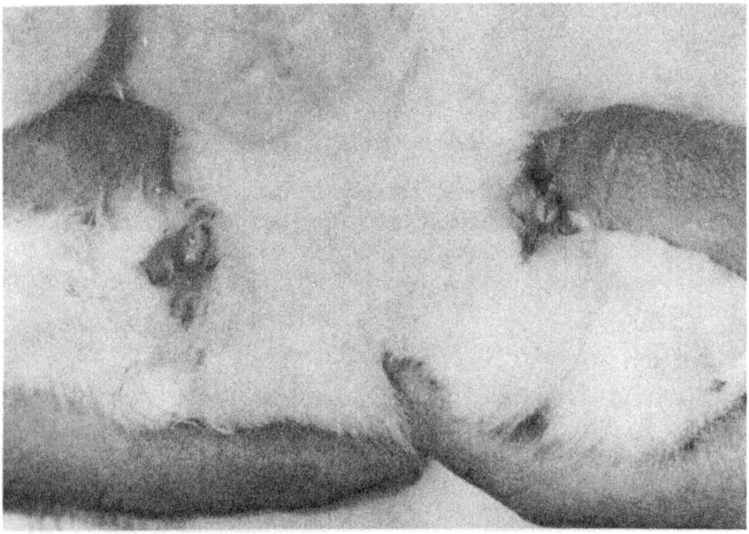

15.6. *Sexing 6-week-old rabbits, male left, female right*

Handling

A rabbit should be lifted by the scruff of the neck with one hand and the rear of the body supported as soon as possible by the other; it is then scooped under the handler's arm, the head region being held firmly against the handler's body with the elbow. Like most animals the rabbit will not struggle if held firmly and securely: an incorrectly held rabbit may lash out and break its back or inflict serious scratches on the handler.

Identification

Rabbits can be marked by ear tattoo, ear tags, or leg bands (Fig A11). As young rabbits grow relatively quickly, it is important to select the correct size of leg bands to avoid discomfort when mature. The most satisfactory method is tattooing, although with a few dark breeds this

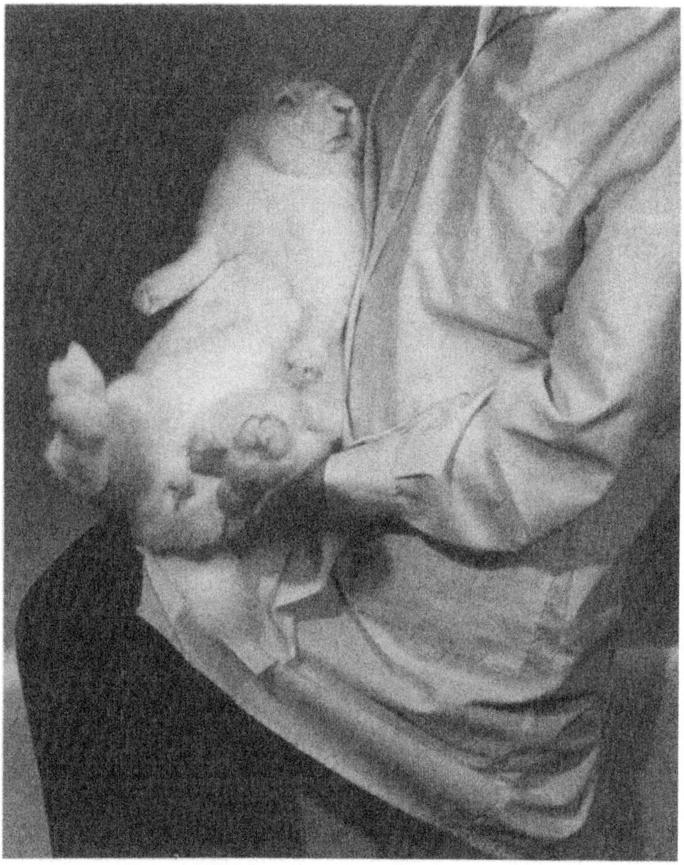

15.7. *Handling adult rabbit for sexing*

may present problems. The majority of breeds, including some with dark fur, have light skins and black tattooing pastes show up clearly.

Cages and bedding

Rabbits are usually kept in cages with wire floors through which faeces can fall. The floors may be welded mesh, usually 17 mm (¾ in), rods or aluminium sheet with punched holes. There is some controversy concerning optimum cage sizes but stock cages of 46 cm × 46 cm × 46 cm (18 in × 18 in × 18 in) and breeding cages of 92 cm × 46 cm × 46 cm (36 in × 18 in × 18 in) appear adequate (Fig 15.9). For breeding females nest boxes containing hay and softwood shavings are provided a few days before parturition (Fig 15.10).

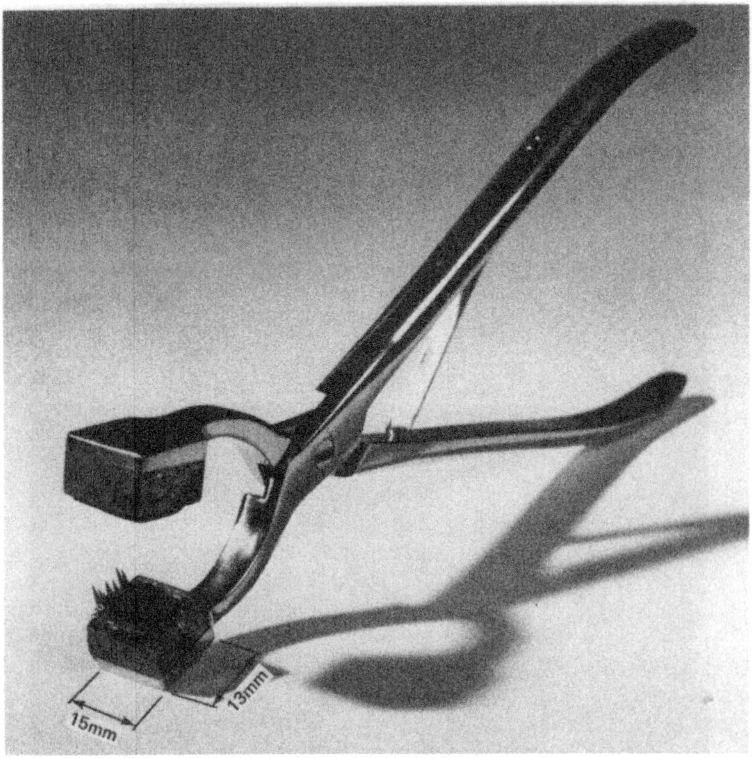

15.8. *Tattooing forceps suitable for rabbit or guinea -pig*

Disease

The important diseases are:

Type	Common name	Causal agent
Bacterial	'Snuffles' or infectious nasal catarrh	*Bordetella bronchiseptica* *Pasteurella septica*
	Syphilis	*Treponema cuniculi*
Viral	Myxomatosis	*Virus*
Endoparasitic protozoan		*Eimeria stediae* (liver)
	Coccidiosis	*Eimeria perforans* } (intestine) *Eimeria magna*
Ectoparasitic mite	Ear canker	*Chorioptes cuniculi* *Psoroptes communis*
	Mucoid enteritis	Obscure, may be coccidia

15.9. *Anodised aluminium rabbit cage with perforated sheet floor*

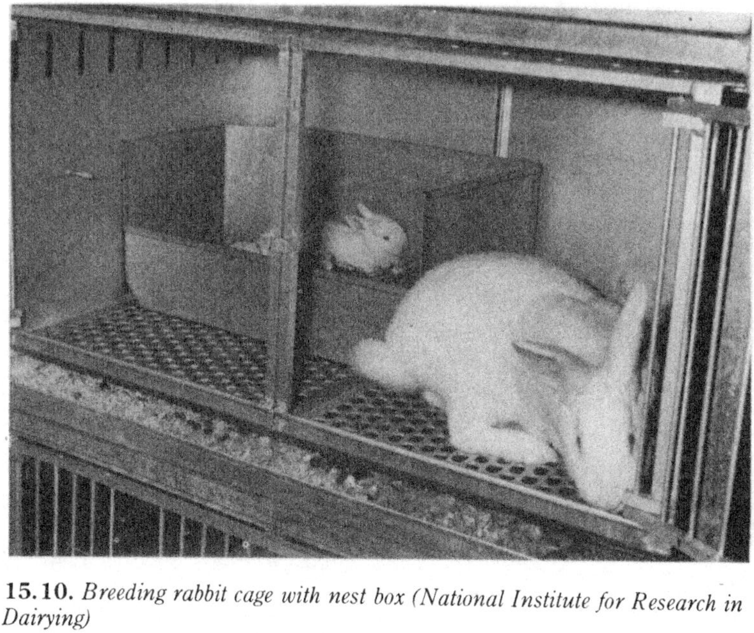

15.10. *Breeding rabbit cage with nest box (National Institute for Research in Dairying)*

The ferret (*Mustela putorius furo L.*) 16

Body temperature	38.8°C (101.8°F)
Age at sexual maturity	9 months
Age at first mating	12 months
Oestrous cycle	continuous
Duration of oestrus	Until mated or end of breeding season
Gestation	42 days
Litter size	6–10
Weaning age	8 weeks

This species, a carnivore related to the otter, stoat and weasel, may be a brown and cream wild type known as a polecat ferret and an albino variety, believed to have been derived from the polecat. The species has been classically used experimentally for the investigation of canine distemper and finds favour today for research into the influenza group of viruses. Their use is increasing with the interest in virology and immunology. It is widely used domestically in catching rabbits. (Fig 16.1).

16.1. *Polecat and albino ferret (National Institute for Research in Dairying)*

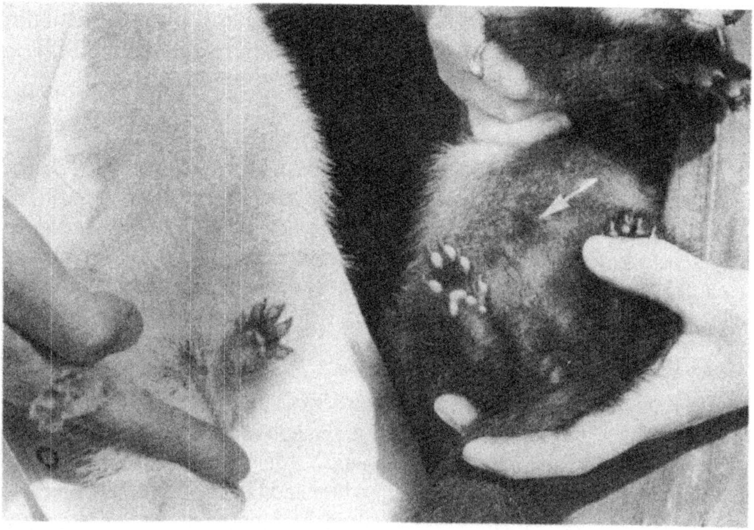

16.2. *Adult female and male ferret, penis arrowed (National Institute for Research in Dairying)*

Anatomy

The ferret is a long, slim carnivore with short legs, the body is very muscular which enables the animal to manoeuvre in confined spaces such as a rabbit burrow.

Reproduction

The ferret starts to become sexually active in mid-winter and is fully so by about March; the male's testes enlarge and the vulva of the female becomes considerably swollen and bright pink in colour at which point the animal is in oestrus. The female will remain in this condition until mating, or if unmated, until the end of the breeding season in August. The condition re-appears again about two weeks after pseudo-pregnancy, abortion or the weaning of a litter. It is possible to obtain two litters during a breeding season and exceptionally three. The breeding season may be extended by the use of an artificial lighting regime (see UFAW Handbook p. 358). During the breeding season both sexes lose weight, up to 50%, though the female will rapidly regain weight, mainly as subcutaneous fat.

When in oestrus the female should be placed in the male's cage for 24 hours and then removed to a cage by herself for five days after which she can be caged with other mated females until a few days before parturition when she should be isolated once more and supplied with plenty of nesting material. When mated females are housed together, however, a close watch should be kept for fighting. Prior to, and for a few days after the birth, the female should be disturbed as little as possible to reduce the chance of her killing the young.

Nutrition

A ferret can be maintained adequately on a diet of two whole dead mice and a third of a pint of milk per day. Canned foods may be used but scouring often results due to lack of roughage. Various pelleted diets have also been used with success. Stock animals may be fasted one day a week, Sundays for convenience, but there should be water available at all times.

Handling

Ferrets have an undeserved reputation for being aggressive, it is more accurate to say that they have a misplaced enthusiasm for investigating objects with their teeth, partly to discover whether they are edible. Ferrets are short-sighted and this adds to their reliance on their teeth. The art of handling strange ferrets is to be quicker than they are: offer the *back* of one hand to them to investigate and then grasp them quickly and firmly round the neck from behind with the other hand. Ferrets, especially males, soon become very tame and may simply be picked up around their shoulders; females should never be entirely trusted, however, especially when they have young.

16.3. *Restraining a ferret (National Institute for Research in Dairying)*

Identification

Albino and lightly coloured polecats may be most easily marked with dye or felt marker. Polecats may be distinguished by their natural markings if there are relatively few animals kept. A non-expandable cat collar with numbered disc may also be used.

Environment

Adult ferrets tolerate cold conditions well, either outdoors or in unheated buildings but breeding animals should be kept at temperatures above 15°C for the benefit of the young.

Caging and bedding

Plenty of bedding should be provided for littering females. The older type wire, solid floored rabbit cage is ideal for holding one or two ferrets, sawdust or shavings being used as litter. The lower 10 cm (4 in) of the cage should be panelled with sheet metal to prevent this litter being rapidly scratched out of the cage. Ferrets may be kept in large groups running free on the floor of a suitably barriered room.

Diseases of the ferret

Type	Common name	Causal agent
Bacterial	Neck and face abscesses	*Staphylococcus pyogenes*
	Tuberculosis	*Mycobacterium tuberculosis* of the human, avian and bovine types
Viral	Canine distemper	Canine distemper virus
	Influenza	Human influenza virus
Ectoparasitic mite	Ear mite	*Otodectes cynotis*

The cat (*Felis catus*) 17

Body temperature	38.6°C (101.6°F)
Age at sexual maturity	8–9 months
Mating weight	2500 g
Oestrous cycle	14–21 days
Duration of oestrus	3–6 days
Gestation	63 days
Litter size	3–6 days
Weaning age	5–6 weeks

The domestic cat, *Felis catus*, belongs to the same family as lions and tigers. There are many breeds of domestic cat, which may be subdivided into short and long haired varieties. Common types are Tabby, Black and Tortoiseshell. Cats bred for research are used mainly in pharmacological, physiological and behavioural studies.

Anatomy

Cats are carnivores and, like dogs, need sharp teeth for tearing and cutting flesh; the canine teeth are enlarged for tearing and the molars are pointed for cutting (Fig 17.1).

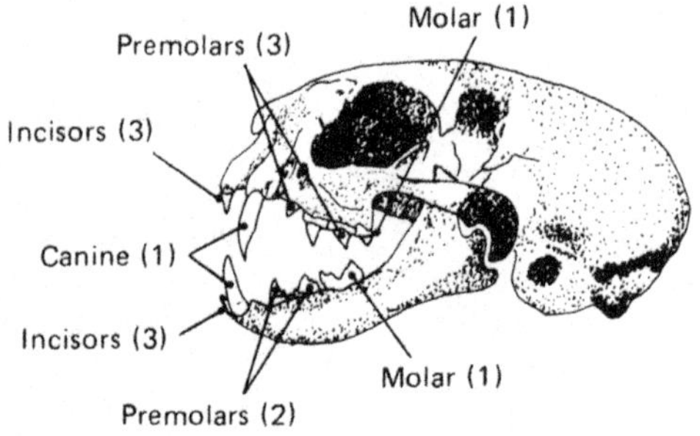

17.1. *Cat skull*

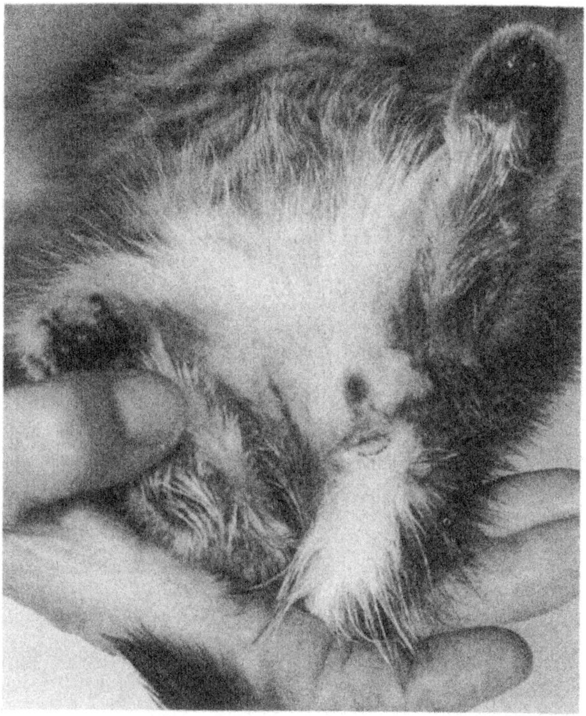

17.2. *Sexing a 6-week-old female kitten*

Cats use their limbs for running, climbing, defence and for obtaining food. Strong toes equipped with sharp retractable claws grip the flesh of an enemy or prey. Dewclaws are not often apparent in cats, but if present can cause problems if they become caught in loose wire.

Reproduction

The cat has no definite breeding season and is polyoestrous, that is it comes into oestrus every 14 days for a period of 3–6 days. A female (queen) on heat will often call continually for the male (tom) and will exhibit rolling and curling up on the floor. When she is ready to accept the male she will arch her back and hind limbs in a distinctive way in readiness for mating (lordosis). The female will allow mating several times until the end of the oestrous period; usually the female is taken to the male's cage for mating. Sexual maturity is reached at about 8 to 9 months and mating should not be attempted before this age.

The gestation period is 60–63 days and fetuses may be detected by palpation after 21 days. Pseudopregnancy is not common in cats, but if it

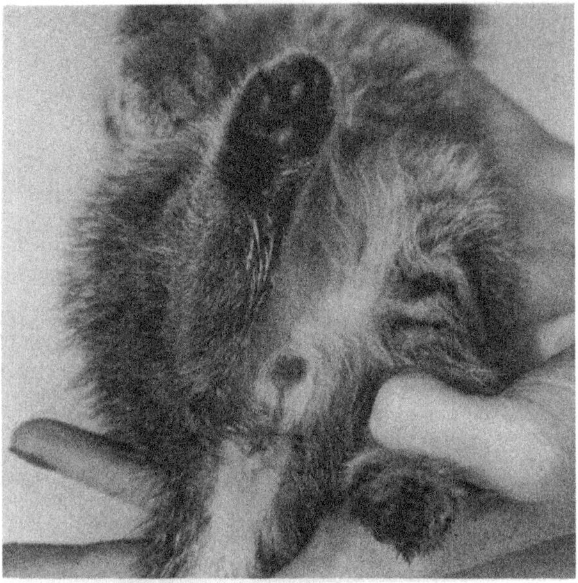

17.3. *Sexing a 6-week-old male kitten*

occurs lasts about 23 days. Ten days before the queen is due to litter she should be placed in a littering (kittening) cage or simply moved into a quiet area. A cat prefers to litter in a dark corner and the provision of a cardboard or wooden nestbox lined with paper or woodwool reduces kitten mortality. Cats do not generally have difficulties with parturition, but it is advisable to have a veterinary surgeon available to deal with any problems that arise.

The female will wash the newborn kittens and bite through the umbilical cord and the kittens will begin to suck almost immediately. It is possible to rear under-nourished or orphan kittens by hand-feeding or by fostering, but careful consideration should be given to the composition of cats' milk so that a suitable formula is used for a milk substitute. Weaning takes place at 5–6 weeks and by then the kittens should be feeding on a solid diet such as tinned meat with the addition of a milk-based food such as 'Farex' (Glaxo Ltd. Greenford, Middlesex).

Nutrition

Cats may be fed a variety of foods, fresh, tinned or dried, and most cats enjoy any of these. Fresh foods such as fish and meat have the disadvantage that they must be cooked before being fed in order to destroy pathogens. Tinned meat or fish is far easier to prepare and is more hygienic, having undergone quality control tests before being marketed. Proprietary diets such as tinned meat, biscuits and expanded diets are therefore straightforward to administer. Dry diets cause the cat to drink more and water should, therefore, be available at all times. Dry diets have a shorter shelf life than tinned foods. Food and water utensils should be cleaned and sterilised daily.

Handling

When attempting to handle a cat that may be nervous and defensive a calm and confident approach is essential. The cat should be stroked and talked to until its confidence in the handler is apparent. When the cat appears calm, lightly grip the scruff of the neck with one hand and place the other hand under the thorax and lift the cat into a secure handling position. For carrying short distances, providing the animal is calm, it is less distressing for it to be tucked under the arm than put in a box. When manipulation for special techniques such as withdrawal of body fluids is required, cats may be wrapped in a thick towel to immobilise them.

Identification

Cats are usually identified by the colour and pattern of their fur, or if they are housed singly, by a cage label. Expanding collars may also be used but these must be adjusted as the kitten grows and must be 'chew-proof'. The danger of using non-expanding collars is that they may become caught in loose fittings and consequently injure the cats.

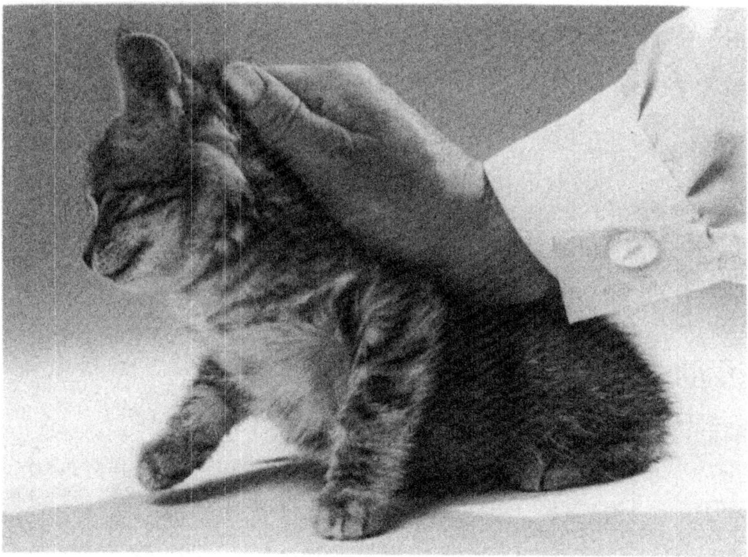

17.4. *Restraining a kitten*

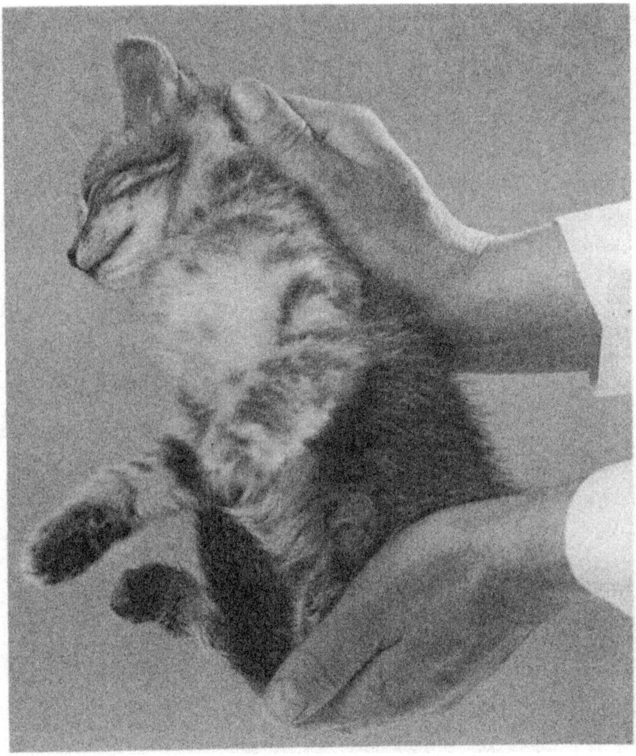

17.5. *Lifting a kitten*

Environment

The cat requires a well-ventilated, draught-proof environment, with a temperature range of 15–18°C (60–65°F) and a relative humidity of 45–55%. Heating should be regulated to maintain a constant temperature within the specified range. Artificial lighting controlled by a time switch to provide 14 hours of light in each 24 hours allows breeding throughout winter and summer.

Caging and bedding

Laboratory cats may be housed in cages or in pens inside a room. They may also be successfully housed in colonies, therefore large areas can be wired off to make pens or good sized rooms can be used for this purpose. The room must be escape-proof, preferably with a double door system, and cages should have cat proof fastenings on the doors because cats are extremely dextrous creatures and may easily open cage doors.

Pens and cages should be provided with litter or dirt trays filled with absorbent material such as sawdust, peat or attapulgite clay ('sterolit') in which the cats may scratch to cover their urine and faeces. Windows that are going to be opened for ventilation must be covered with wire netting to prevent the cats from escaping and from scratching the window frames.

Walls and floors should be non porous, hard wearing, non slip and easily cleaned. An important point to consider if drains are provided is that the floor should slope towards them to assist with disposal of excess water and effluent. Both asphalt and epoxy resin floors are suitable for catteries; tiles are preferable but are expensive and difficult to seal effectively. Only materials which are easily cleaned should be used in catteries; these include metal and plastic.

An important item is a wooden scratching post on which the cats can sharpen their claws; this may be part of a tree or purpose made.

Cat pens may be designed with exercise runs accessible via a swinging flap. Pens should be separated by solid walls to prevent groups of cats coming into contact and to reduce cross-infection. These pens provide plenty of space for exercise and with thoughtful planning a more natural environment can be created.

Common diseases

Type	Common name	Causal agent
Bacterial	Septicaemia	*Streptococcus haemolyticus*
	Pneumonia	*Bordetella bronchiseptica*
Viral	Feline panleucopaenia	DNA type virus
	Feline pneumonitis	Herpes type virus
Endoparasitic		
cestode	Tapeworm	*Taenia pisiformis*
		Taenia taeniaformis
		Dipylidium caninum
nematode	Roundworm	*Toxascaris leonina*
		Toxascaris mystax
Ectoparasitic		
mite	Ear canker	*Otodectes cynotis*
	Body mange	*Notoedres cati*
		Demodex folliculorum
Fungal	Ringworm	*Trichophyton felineum*
		Microsporum canis
Physical	Furballs	Accidental ingestion of hair
Physiological	Urinary calculi	Build up of mineral salts

The dog (*Canis familiaris*) 18

Body temperature	38–38.7°C (100.4–101.6°F)
Age at sexual maturity	9–12 months
Age at first mating	14 months
Oestrous cycle	Two seasons per year
Duration of oestrus	7–14 days
Gestation	60–63 days
Litter size	4–8
Weaning age	6–8 weeks

There are many breeds of dog ranging in size from toy varieties like the chihuahua to large working dogs such as the St Bernard, and their use in domestic life is considerable (such as breeding, showing and working).

The use of dogs for research requires that they are of uniform size and have a good temperament making them easy to handle. For these reasons the beagle is a most popular model but labradors, greyhounds and mongrels are also used.

Anatomy

The dog has two consecutive sets of teeth. The first set are called deciduous or milk teeth; these are displaced by the second or permanent teeth (Fig 18.1).

The dog is carnivorous and its teeth are developed to allow it to eat and prepare the food for digestion. Incisors are used for tearing and cutting food and the molars are used for grinding. In a healthy dog the teeth should be clean and free of tartar and be firmly set in the jaw.

Dogs have four toes and claws on each foot. At the back of the foot dew claws grow which do not have a function, with the exception of the Pyrenean mountain dog which is said to use them for climbing in snow. Dew claws are usually removed soon after birth to prevent the dog from causing accidental injury to itself or to other dogs. Some breeds, such as spaniels, have their tails docked; this procedure should only be performed by a veterinary surgeon.

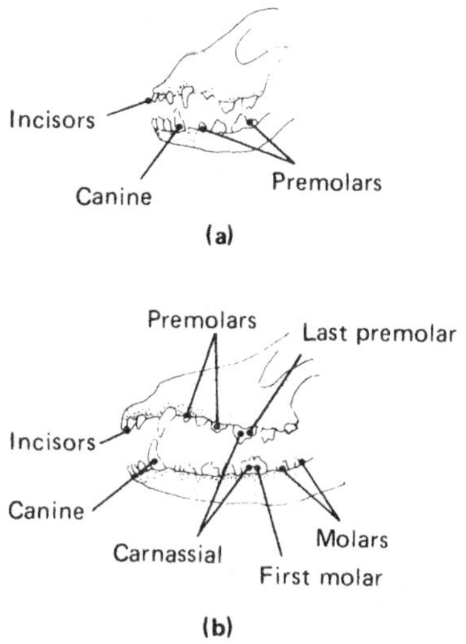

18.1. *Dentition of dog (a) immature (b) adult*

Reproduction

Usually there are two breeding seasons per year and a season may occur as early as six months of age in some bitches. The vulva of the bitch begins to swell just before the onset of oestrus, followed by a discharge of blood and mucus. One week later the discharge appears straw coloured and the bitch should be ready for mating about the eighth day after the onset of oestrus. A repeat mating within two or three days is advisable to ensure a successful mating. After twelve to fourteen days the cycle finishes and the bitch will not mate until her next season; the length of the cycle may vary between breeds.

Ten days before parturition the bitch should be placed in a whelping pen containing warm bedding material such as soft wood chippings for the puppies to nest in. An infra-red lamp suspended 60 cm above the whelping area will provide extra warmth for the puppies when they are born. A veterinary surgeon should be available to assist with difficult births.

After the birth of the puppies the bitch will lick them clean and within a short time they should be sucking. An inspection a few days after birth is advisable to check for deformed or weak puppies and infected navels. Dew claws should be removed at the time of the inspection.

Nutrition

Several proprietary dog foods are available; some laboratories prefer to feed a mixture of tinned meat and biscuits, others feed expanded biscuit diet alone. The expanded biscuits are formulated to provide a balanced diet for maintenance and breeding without the need for supplements. Biscuits may be fed *ad libitum* from a food hopper to prevent contamination. Pregnancy and lactation are conditions where a bitch's health may deteriorate if the correct attention to diet is not forthcoming, therefore additional food and vitamin supplements are essential to maintain the health of the bitch and her puppies.

Puppies can rely on their mother's milk until they reach about four weeks of age, when they are encouraged to become independent and eat soft milky foods such as 'Complan' (Glaxo Ltd. Greenford, Middlesex). Meal or biscuits suitable for puppies, soaked in water or milk, will encourage them to accept solid food gradually.

Water must be available at all times, preferably administered from an automatic watering system to avoid contamination.

The question of giving bones is debatable. When dogs are housed in groups, feeding bones may cause fighting, depending on the nature of the dogs and the system of housing. Many laboratory workers do not feed bones and have seen few dental problems as a result. If bones are fed they should be selected to avoid those that splinter and cause dogs to choke, for example those from chickens, rabbits and pigs.

Handling

The dog should be approached quietly and confidently; talking will help to calm and reassure the animal. Approaching the dog from the front, hold the scruff of the neck with one hand, at the same time slip the other hand under the body to support the dog when lifting. It is not always necessary to hold the scruff of the neck, and a good handler who is familiar with his animals may not need to use this method. If the dog is nervous and resents any attempts to be handled, a dog noose consisting of a piece of rope forming a loop, and running through a tubular handle, can be placed over the dog's head and neck and gently tightened. If close examination is necessary, the dog may need to be muzzled by placing a 7.5 cm (3 in) width of bandage round the nose and jaw and tying it behind its head.

Identification

The most effective and permanent method of identification is tattooing. The part of the ear to be tattooed is shaved with a scalpel blade, and smeared with antiseptic tattooing paste, the appropriate code is inserted into the plier and punched into the ear, taking care to avoid piercing major blood vessels.

Collars carrying a numbered disc are used but can cause serious problems: they may get caught in loose wire and cause injuries, the chain may break and be lost or simply cause discomfort.

Environment

The recommended temperature is 13 – 18°C (55 – 65°F) with a relative humidity of 45–55%. The dog appears to withstand cold more easily than hot conditions and so every effort should be made to maintain the correct temperature for its comfort. Controlled lighting is not essential for breeding.

Cages and bedding

Dogs can be housed individually or in groups depending on the space available. Housing that provides warm sleeping facilities combined with an outside run for exercise is preferable to one area used for living and sleeping. The sleeping area should have a bed board raised at least 45 cm off the ground or a draught free insulated area on the floor that will keep the dogs warm.

The outside run should be constructed with a hardwearing surface such as concrete which should slope for good drainage to facilitate daily hosing. A swinging door flap in the wall will allow access to the sleeping area. Whelping bitches will require a bed of soft woodchips or any warm, absorbent material that will keep the puppies warm and dry without suffocating them.

Important diseases

Type	Common name	Causal agent
Bacterial spirochaete	Jaundice	*Leptospira canicola*
		Leptospira icterohaemorrhagiae
Viral	Distemper	Virus
	Hard pad	Virus
	Infectious hepatitis	Virus
	Parvo virus infection	A corno virus
	Rabies	A neurotropic virus
Endoparasitic nematode	Roundworm	*Toxacara canis*
		Toxascaris leonina
Endoparasitic cestode	Tapeworm	*Taenia pisiformis*
		Taenia hydatigena
		Dipylidium caninum
Ectoparasitic mite	Ear canker	*Sarcoptes scabiei*
		Otodectes cynotis
	Body mange	*Sarcoptes scabiei*
		Demodex canis

Simians

19

Physiological data have been omitted due to the wide variation between simian species.

Anatomy

The simians (non-human primates) generally resemble humans in their anatomy; forward facing eyes, hands with a gripping thumb and a reproductive system having a menstrual cycle rather than an oestrous cycle. The many species are categorised according to their geographical distribution, Old World (Rhesus, Cyanomolgus, and Baboons), New World (Marmosets and Squirrel monkeys). Most New World monkeys have prehensile tails (an exception is the marmoset), the Old World monkeys do not. Simian dentition is similar to human but the incisors and canines are often much enlarged.

Reproduction

The simian exhibits a monthly menstrual cycle and usually mates at the middle of the cycle. There is a tendency towards spring births. The period of oestrus lasts about 7–14 days and during this time the female seeks to mate with the male. The cycle lasts approximately 28 days but varies from 18 days (capuchin) to 35 days (baboon and chimpanzee). In some species of apes and monkeys the skin around the genitalia of the female takes on a vivid colour, reaching maximum at ovulation after which it gradually returns to normal. It is thought to indicate to the male that the female is sexually responsive.

Nutrition

Complete diets for primates are available which eliminate the need to offer fresh fruit as a source of vitamin C; however, the feeding of fresh fruit, vegetables and peanuts relieves boredom in the colony. Monkeys waste large amounts of food and some restriction of the quantity offered is advisable; uneaten food should be removed from the cage daily. Vitamin D_3 should be included in the diet of New World monkeys as they are unable to utilise vitamin D_2.

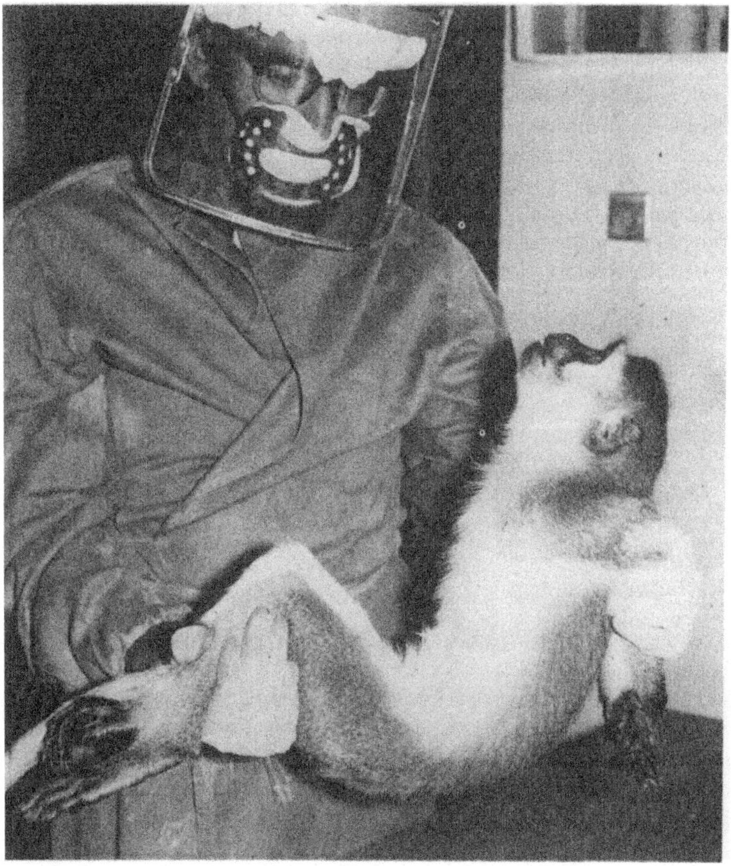

19.1. *Technician wearing full protective clothing whilst handling sedated vervet monkey*

Handling

With the exception of New World monkeys which are small and reasonably easily restrained, simians should never be handled unsedated and even then they should be treated with caution and protective clothing should be worn. Where repeated anaesthesia would interfere with the experiment, a suitably designed 'crush' cage allows an unsedated monkey to be indirectly controlled; only the required limb need be handled by the experimenter for injection. Apparently healthy monkeys may be sub-clinically infected with various human pathogens, for example *Shigella spp*, *Salmonella spp* and *Mycobacteria spp*. Newly imported monkeys may carry rabies, B virus or Marburg virus. These three agents are generally fatal to humans. The minor inconvenience of wearing suitable protective clothing can never be equated to the risks involved if these precautions are not taken. Application of the *Health and Safety at Work Act 1974* makes these precautions obligatory. (Readers are strongly recommended to consult *Hazards of Handling Simians*, Ed. Perkins and O'Donoghue, 1969, Laboratory Animals, London.) Technicians are liable to be spat on, coughed and urinated over and can be the targets for well-aimed faeces, food and cage debris. The following clothing is desirable and should only be removed from the primate unit after being sealed in bags for sterilisation:

Underclothes.
Full-length robe with non-gaping fastening and which is long enough to cover the boots or, long-sleeved, round-neck T shirt and trousers that may be worn over the outside of the boots.
Disposable face mask, full face visor and a hood which completely covers the head and shoulders except for the face.
Disposable surgeons' gloves worn with the wrist section covering the ends of the sleeves of the gown or T shirt.
Wellington boots.

As a further safeguard, a shower should be taken on leaving the unit.

When sedating a monkey there should always be two people present in case the animal recovers and escapes. For similar reasons, monkeys should never be left on benches or in unsecured cages whilst recovering from the effects of anaesthesia or sedation. The recovery rate varies and may be very rapid. The precautions described should prevent the escape of a monkey in the unit but nets and a catching cage should always be available in case of accident. With large species a tranquillising-dart gun would be a wise investment.

Identification

Tattooing is the usual method employed, either on the chest, ears or lower lip. The use of a chain with a numbered metal disc around the waist has been suggested but is impractical in use because it may become caught in the cage and cause injury to the animal.

Caging

Monkey cages should be solidly constructed to withstand the considerable strength of the occupants. A 'guillotine' door rather than a hinged type allows it to be opened slightly for the introduction of fruit without the risk of an animal escaping. The door and removable floor should be padlocked in position; monkeys soon learn to open conventional catches. A bench or rail is essential for the monkey to sit on to prevent cage paralysis caused by the unnatural posture if the animal has to sit on the floor.

A false back to the cage which can be slid to the front 'crushing' the animal against the front bars is usually provided; this permits injection of a drug or sedative into the monkey between the bars so avoiding the need to handle the animal.

Environment

New World monkeys need to be kept at elevated temperatures (27°C, 80°F) to match that of their native climate but Old World monkeys are able to adapt to the British climate successfully; some Rhesus monkey units have outdoor runs which are used by the animals all the year round.

Common diseases

Type	Common name	Causal agent
Bacterial	Dysentery	*Shigella flexneri, sonnei* & *schmitzei*
	Tuberculosis	*Mycobacterium tuberculosis* of the human, avian and bovine types
	Pneumonia	*Diplococcus pneumoniae* *Klebsiella pneumoniae*
Viral	Rabies	A neurotropic virus
	Virus B	*Herpes virus simiae*
	Marburg or Green Monkey Disease	Thought to be a virus but not isolated
Endoparasitic nematode	Roundworm	*Oesophagostumum opiostumum*

The Chicken *(Gallus domesticus)* 20

Body temperature	40.5°C (105°F)
Age of sexual maturity	5–6 months
Incubation	21 days

Most chickens that are bred commercially for meat or eggs are hybrids specially developed for maximum efficiency. Few commercial chicken farmers keep the traditional breeds from which these hybrids have been derived. Breeds are often classified by weight, for example: light (White Leghorn, 1½–2½ kg) and heavy (Rhode Island Red and Light Sussex, 2–2½ kg). A fast growth rate, particularly in modern broiler hybrids, makes the chicken a useful subject for nutritional studies where a rapid response to dietary effects is required. Chicken eggs are widely used in virological studies.

Anatomy

There are several features of anatomy that are peculiar to birds. The red blood cells are nucleated, the bones are hollow and there are air sacs in the body cavity which form extensions of the lungs thus increasing the tidal volume. The gut is short and terminates at the cloaca which is the common opening for reproductive and excretory systems, (Fig 20.1). Sexing day old chicks is an important skill in a commercial hatchery; in some hybrid strains feather colour is sex linked which simplifies the procedure.

Reproduction

Young hens (pullets) start to lay at 5–6 months and if fertile eggs for hatching are required cockerels should be introduced several weeks earlier; 6–9 cockerels per 100 females are used depending on the breed. The female stores semen and will lay fertile eggs over a period of several days without further matings. Fertile eggs can be stored for 10 days when kept at 13–16°C (55–61°F) at a relative humidity of 80–90% and should be turned at least twice per day. In nature a broody hen

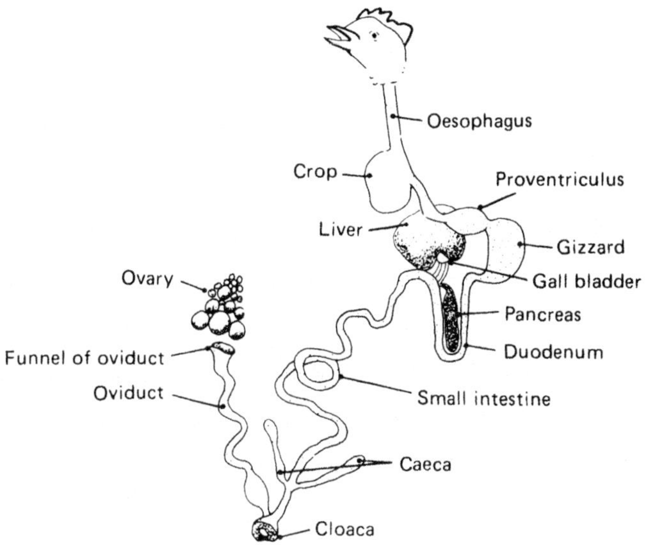

20.1. *Hen alimentary canal and reproductive system*

would sit on 10–12 eggs for 21 days. Commercially most eggs are hatched in an incubator. Incubators are either the still air flat type or the forced draught cabinet type: their features are listed below.

	Flat Type	**Cabinet Type**
Setting/hatching	Combined	Separate
Egg position	Horizontal	Vertical – broad end up
Turning	Rolled 180°	Tilted 90°–45° each side of vertical
Ventilation	Manually operated vents	Mechanical
Hatching conditions	1–21 days 38.4°C (101.5°F) RH 60%	1–18 days 37–37.8°C (98.6–99.7°F) 18–21 days 36.9–37.5°C (98.4–99.5°F) RH 60%

Several factors must be born in mind to achieve good hatching results. Eggs should be collected frequently. Dirty, cracked, misshapen and abnormally large or small eggs should be discarded. The correct storage conditions should be maintained. To prevent a drop in temperature when additional eggs are introduced to the incubator they should first be warmed. Incubators should be thoroughly disinfected and fumigated between hatches.

Nutrition

A wide range of proprietary diets presented as pellets, crumbs or mash is available for poultry of all ages. Diets are classified as high energy (over 12.0 MJ of metabolisable energy/kg), medium energy (11.0–12.0 MJ of ME/kg) or low energy (10.6–11.0 MJ of ME/kg). High energy diets are required for light hybrid layers (1.5 kg) and for broilers which are to reach about 1.8 kg in 56 days. For heavy birds and for rearing, high energy diets can be both wasteful and harmful. Laying hens have a greater requirement for protein; an egg contains about 14% protein. Chicks up to 6 weeks of age require 20% protein and thereafter 18%. Protein quality is important, particularly the levels of the amino acids lysine and methionine.

Because of their fast growth rate chickens soon exhibit clinical symptoms if a diet is deficient in a particular vitamin or mineral.

Vitamin/Mineral	Deficiency symptoms
Riboflavin B_2	curled toe paralysis
Vitamin K	prolonged blood clotting time
Vitamin D	rickets
Vitamin E	nervous disorders (crazy chick disease)
Pantothenic acid	dermatitis and poor feather formation
Vitamin B_{12} (cyanocobalamin)	abnormal red blood corpuscles
Calcium and phosphorus	malformed bones
Manganese	weak skull, perosis (slipped tendon)

Water is essential; a day old chick may drink up to 30 ml/day whilst an adult may drink 180 ml/day. Water can be supplied from nipple drinkers, self-filling bowls or open troughs.

Handling

Chicken are more easily caught in dimly lit rooms; young birds can be grasped around the body, older birds can be caught by the legs. When carrying a chicken the body should be held under the arm thus trapping the wings, and the legs held firmly in one hand (Fig 20.2).

Identification

Coloured or numbered wing tags (e.g. 'Ketchum' tags) or leg rings can be used (Appendix 1 Fig A11). Care must be taken when applying these as the bird's fast growth may cause tags or rings to become too tight.

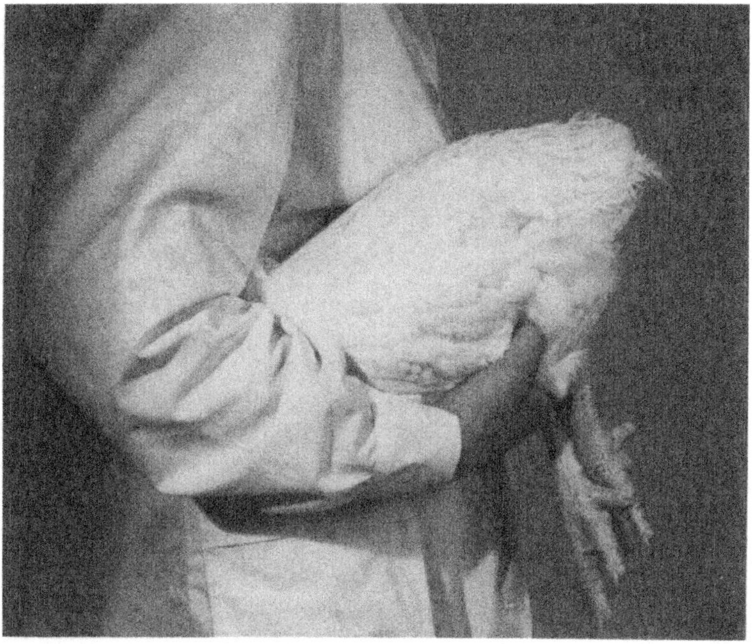

20.2. *Carrying a chicken*

Housing

Adult birds on experiment are usually kept in cages. As a result of the report of the Brambell Committee in 1965, Codes of Practice for chicken housing on farms have been produced by the Ministry of Agriculture, Fisheries and Food, but it is unlikely that experimental birds would ever be kept in such cramped conditions as the minimum areas suggested. Almost any large cage with a wire floor will be suitable for chickens although rabbit cages may not be high enough for heavy birds.

Commercially, large groups of birds are reared in deep-litter houses. Laying birds are usually housed in battery cages.

In the laboratory chicks are more conveniently reared in tier brooders (Fig 20.3). Each tier has its own heat source, usually a heated pad under which the chicks can gather. As they grow the numbers in each section may be reduced if necessary to avoid overcrowding.

An alternative to brooder rearing is a small hot room where chicks are kept in cages and the whole room is heated to a brooding temperature of 32–35°C (90–95°F) and reduced by 3°C per week for three weeks until a temperature of around 18–21°C (65–70°F) is reached.

20.3. *3-tier chick brooder (National Institute for Research in Dairying)*

The relative humidity should be kept at about 50%. Various lighting regimes are used; day old chicks are usually started on 18 hours light and 6 hours dark, and other regimes are used, depending on management, at a later age to induce laying, broilers are usually kept on 23 hours light, 1 hour dark.

Common diseases

Type	Common name	Causal agent
Bacterial	Salmonellosis	*Salmonella spp.*
	Chronic respiratory disease (CRD)	*Mycoplasma spp.*
Viral	Infectious laryngotracheitis (ILT)	
	Newcastle disease (fowl pest)	
	Fowl pox	
	Avian leukosis	
	Marek's Disease	
	Avian encephalomylitis	

Type	Common name	Causal agent
Endoparasitic		
protozoan	Coccidiosis	*Eimeria spp.*
nematode	Large roundworm	*Ascaridia lineata*
	Gapeworm	*Syngamus trachea*
cestode	Tapeworm	*Rallietina spp. Hymenolepis spp.*
Ectoparasitic		
louse	Body louse	*Menacanthus stramineus*
	Shaft louse	*Menopon gallinea*
	Head louse	*Cyclogaster heterographus*
	Wing louse	*Lipeurus caponis*
mite	Red mite	*Dermanyssus gallinae*

Disease 21

General principles

Many conditions may make an animal ill such as lack of water or food, inadequate diet, stress, physical deformity, disease or infection, physical damage and old age.

A **disease** is a condition where part or parts of an animal body cease to function normally.

An **infection** is a disease resulting from contamination by a disease-producing (pathogenic) organism. Signs of the infection may not always be apparent. To establish whether an animal is ill and to find symptoms, an animal is given a clinical examination. To avoid missing points this should be carried out logically. Firstly, observe the animal and its cage for any abnormalities such as unusual posture or blood-stains in the cage. Secondly, remove the animal from its cage or pen and examine it carefully from head to tail. Signs of ill health may be apparent from the abnormal condition of the nose, mouth, teeth, eyes, ears, fur, abdomen, feet, legs and ano-genital region and from abnormal body temperature.

When animal tissue is damaged, whatever the cause, inflammation results. Tissue can be irritated by infection with bacteria, viruses, protozoa, fungi, ectoparasite infestation, accidental trauma or chemical and physical damage. Bacteria can settle and multiply on intact body surfaces without causing inflammation (commensalism). Many are found in normal animals, for example in the gut. If sufficient micro-organisms invade a susceptible tissue and multiply, the result is infection. The integrity of the surface layer is the most important local defence mechanism. Other defence mechanisms are the production of tears and mucus, high acidity of gastric and vaginal fluids, activation of the host's immune system and the action of cilia, for example, in the respiratory tract.

Once the defences are breached other factors determine whether the organisms will initiate infection:

- whether or not they are species specific

- how numerous they are
- portal of entry, for example a break in the skin is more likely to be infected with *Streptococci* or *Staphylococci spp.* whereas the gut is more likely to be infected by ingested *Salmonellae*
- receptivity of the host, for example, age or stress.

Once the infection is established the pathogen manufactures harmful substances. These may be endotoxins which are retained within the pathogen and are only released into tissue when it dies; or exotoxins which diffuse readily from the pathogen and can exert their influence away from it or from the portal of entry. For example, an exotoxin from *Clostridia tetani* affects the central nervous system not the skin where the organism entered the body. If the toxin is very irritating, acute inflammation is the result. Chronic inflammation will result from a less violent but persistent toxin. It is important to remember with respect to disease or infection, that acute means short, sharp and sometimes violent, whereas chronic means less violent, long lasting and persistent. Both chronic and acute disease can be fatal.

A characteristic response to bacterial toxins is a raised temperature or fever. Excessively high temperature may depress antibody production or even destroy antibodies already formed. It may also cause breakdown of tissue protein and thus loss of weight, degeneration of hair and general tissue depletion. An abnormal temperature may also be the result of other pathological conditions such as heat-stroke or poisoning.

In order to prevent an animal from contracting a disease it should ideally be prevented from coming into contact with the pathogen responsible. The obvious way of doing this is to operate a barrier as described in Chapter 10. With larger animals it may be economically sound to vaccinate against specific diseases but this is usually not so with the smaller species unless their value increases dramatically during the course of the experiment. Vaccination may invalidate experimental results.

Vaccination is the practice of infecting an animal with either dead or attenuated (weak) pathogens. Attenuation is the technique of rendering pathogens harmless by various ways, such as passage, heat or chemical treatment. The vaccine will cause the animal to produce its own antibodies against the pathogen and these will provide protection if challenged by the normal pathogen at a later date. The immunity achieved lasts for a variable time depending on the organisms involved. In some cases a booster vaccination is required at regular intervals to maintain the level of antibodies and therefore the protection. Common examples when protection by vaccination is used are hardpad and distemper in dogs, feline enteritis in cats and myxomatosis in rabbits.

As part of any disease prevention routine, records should be kept of

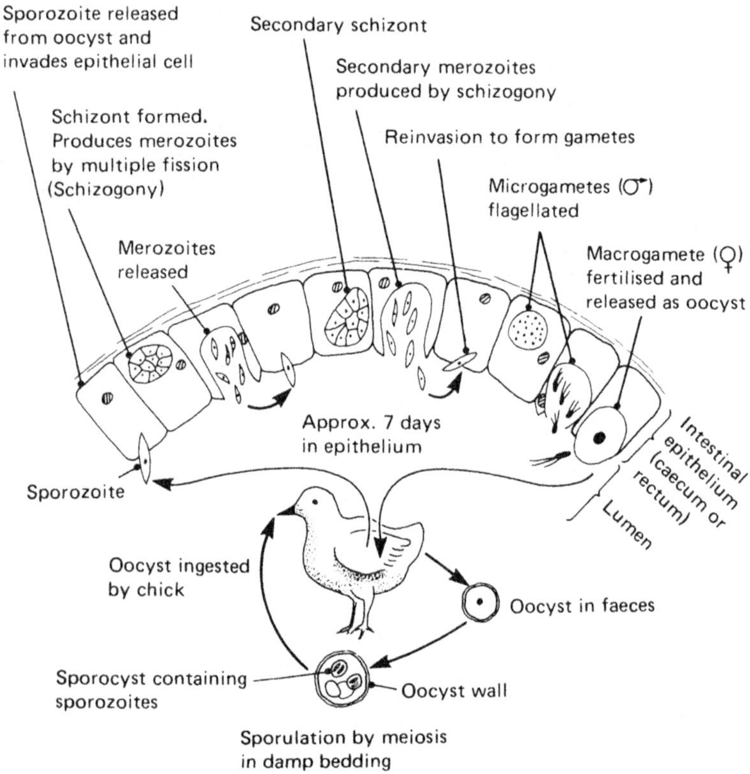

21.1. *The life-cycle of a coccidian protozoan parasite* Eimeria tenella *in chick*

all health problems and deaths. Graphs or a table prepared from these records will highlight epidemics or high mortality and control measures may be taken to prevent spread of the disease.

Unusual or unexplained deaths should always be fully investigated by a pathology laboratory and regular microbiological screening of colonies for evidence of pathogens should also be carried out. If there is not a diagnostic or pathology laboratory attached to the unit the local Ministry of Agriculture, Fisheries and Food Veterinary Investigation Centre will carry out post mortem examinations.

Parasites

Conventionally kept animals may suffer from infestations of external parasites (ectoparasites) and internal parasites (endoparasites). A parasite is an organism which lives temporarily or permanently on another, the host, benefiting from the association and causing harm to the host. All or part of a parasite's environment is formed by another animal and it is metabolically dependant on that animal. Parasites are usually from one of the following groups: protozoans, platyhelminths (flukes and tapeworms), nematodes (roundworms), ticks, crustaceans and insects.

Many parasites have evolved complex life cycles with large numbers of progeny and some of these life cycles involve a secondary or intermediate host.

Common protozoan parasites in laboratory animals are coccidia, different species of which can infect rabbits, poultry, guinea-pigs, sheep and goats and may give rise to the disease coccidiosis. The life cycle of the coccidian *Eimeria* does not rely on a secondary or intermediate host (Fig 21.1).

Some endoparasites have complex and sometimes bizarre life cycles often involving a secondary host. A common example is the liver fluke (*Fasciola hepatica*) (Fig 21.2), which may infect small ruminants.

Apart from depriving the host of its own food, endoparasites may cause much tissue damage and in many cases severe infestations will result in the death of the host. Ectoparasites can also cause severe tissue damage usually to the skin although this is rarely fatal. They are a serious health hazard because by carrying pathogens they may form the route of infection for several diseases affecting laboratory animals, for example myxomatosis and leptospirosis. The principal ectoparasites are mites, ticks, lice and fleas.

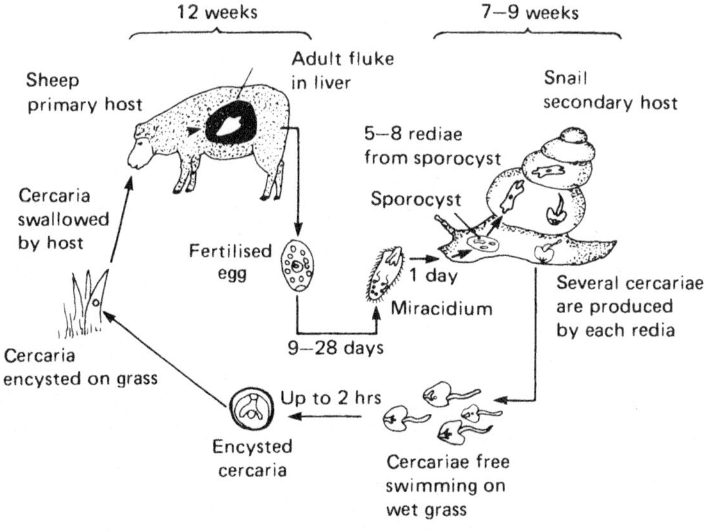

21.2. *The life-cycle of the liver flake* (Fasciola hepatica) *in sheep*

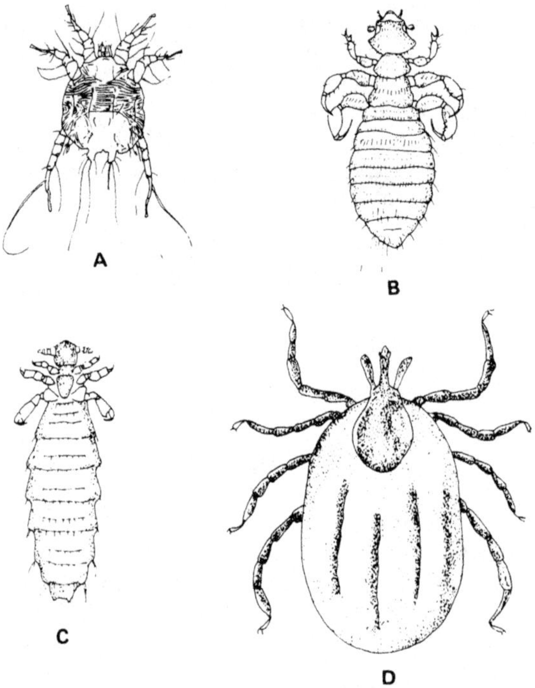

21.3. A *Ear-canker mite* (Psoroptes communis)
B *Guinea-pig louse* (Gyropus ovalis)
C *Rat louse* (Polyplax spinulosa)
D *Castor-bean tick* (Ixodes ricinus)

Mites

There are many different families of mites that affect laboratory animals, causing a variety of conditions including mange and ear canker. Some lay their eggs off the host, particularly those found on birds, and some complete their life cycle on the host. The stages of the life cycle are eggs, six-legged larvae, eight-legged nymph and adult; the cycle may take only one week to complete. Crusty scabs are a familiar sign of mite infestation, these are formed by the dried exudate from the host.

Ticks

Ticks are often described as either hard or soft. Both feed on the blood and tissue fluids of their hosts. The stages of the life cycle are similar to those of mites but the whole process may take from a few weeks to two years. Tick eggs are laid off the host but the six-legged larvae infect a new host for a few days to feed on blood. They then drop off and moult to form nymphs, which may remain quiescent for a long time, for example during autumn and winter, until they invade another host for a short feed. They drop off and moult, this time to form adults.

Lice

Lice can be divided into two sub-orders, the *Anoplura* or sucking lice, and the *Mallophaga* often misnamed as 'biting lice', but which include various feeding types, some of which suck blood. The sticky eggs (often called 'nits') are always attached to the hairs or feathers of the host. Nymphs hatch from the eggs and after three ecdyses or moults become adults, the cycle taking about 30 days. Lice are species specific and therefore do not normally affect other animals.

Fleas

Fleas are parasitic and in adult form, as in the case of lice, are only found on mammals and birds, feeding on the blood of the host by means of specially adapted mouthparts. The eggs, unlike those of lice, are dry and tend to drop off the host. The larvae develop away from the host feeding on the blood-containing faeces of the adult fleas, a necessary factor for their development.

The egg-larva-pupa-adult life cycle takes from eighteen days to several months depending on species and environmental conditions.

The varying life-styles of these ectoparasites must be taken into consideration when planning treatment. Treating the host animal will not be enough as many life cycles involve stages away from the host. Also some stages in the life cycle may be resistant to treatment.

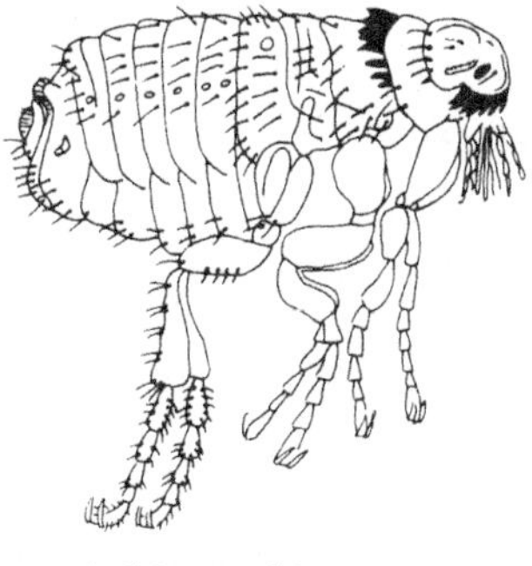

21.4. *Dog flea* (Cetenocephalides canis)

Zoonotic diseases (transmissible between animal and man)

A number of diseases are transmissible between animal and man and therefore if one is working with animals it is advisable to inform the family doctor. Some diseases, particularly those of non-human primates, present a very serious risk and stringent handling procedures have been evolved for this group. Consideration should be given to the *Health and Safety at Work Act* when establishing these handling procedures as employers have an important responsibility to ensure staff are reasonably protected against all hazards.

Some important zoonotic diseases are listed below.

Non-human primates

Tuberculosis
Salmonellosis
Dysentery – *Shigella dysenteria 2*
– *Shigella flexneri 4b*
– *Shigella sonnei*
B-virus – *Herpes simiae*
Marburg virus
Rabies
Hepatitis
Helminths – *Strongyloides*
Protozoans – *Entamoeba*
– *Balantidium*
Toxoplasma

Ungulates

Anthrax
Pox virus
Brucellosis (undulant fever)
Looping ill
Ringworm
Orf
Foot and mouth disease
Erysipelas

Dogs and cats

Hydatidosis
Leptospirosis
Cat-scratch fever
Rabies

Ferrets

Influenza virus

Equidae

Glanders
Influenza

Avians

Newcastle disease
Psittacosis/Ornithosis

Rodents and rabbits

Salmonellosis
Pseudotuberculosis – *Yersinia pseudotuberculosis* (rabbits)
Rat bite fever – *Streptobacillus moniliformis*
Leptospirosis (jaundice)
Lymphocytic choriomeningitis LCM (mice)
Ringworm – *Trichophyton mentagraphytes*
– *Microsporum spp.*
Intestinal parasites – e.g. tapeworm
Hymenolepis nana in mice
Listerium

Diseases of the mouse

Salmonellosis – a bacterial disease

Also known as mouse typhoid this disease is caused by *Salmonella enteridis* or *S. typhimurium*. Infection occurs through food or bedding contaminated by faeces of infected rodents particularly wild rats and mice.

In the acute form there is loss of condition, diarrhoea and conjunctivitis and death within 7–14 days, in the chronic form the animal loses condition but recovers to become a carrier of the disease and may spread infection for an extended period. At post mortem examination there is generalised inflammation of internal tissues, peritonitis, enlarged lymph glands, enlarged spleen and liver, and consolidated lungs.

To prevent the disease a high standard of hygiene is essential, animals must not be overcrowded and it should not be possible for food or bedding to be contaminated by wild rodents. All mice brought into the unit from an outside source should be quarantined away from existing stocks. In the event of an outbreak the room containing the infected animals should be isolated, the animals killed and burnt along with their food and bedding and the room and equipment sterilised.

Tyzzer's disease – a bacterial disease

The causal bacterium *Bacillus piliformis* occurs as an intestinal commensal in many mouse colonies and only under the stress of overcrowding and poor management does the disease become epidemic.

Affected animals lose weight, show signs of diarrhoea, become moribund and may die. At post mortem examination the liver is studded with foci and the gut may be inflamed.

The disease may be prevented by good husbandry and outbreaks controlled with tetracycline ('Terramycin', Pfizer Ltd, Sandwich, Kent) usually administered in the drinking water.

Ectromelia – a viral disease

Also known as mouse pox, the virus may enter abrasions on the feet and tail from food and bedding contaminated by infected mice. Scabs develop on the feet and tail, bald patches appear on the body and the animal may die. In the chronic state scabs form all over the body and the mouse should be killed before tail and feet are sloughed off.

To prevent the disease mice from outside sources should not be mixed with existing stocks. If the disease is suspected the animals should be isolated and a sample sent for laboratory examination. If the disease is confirmed animals should be killed and burnt together with their food and bedding. The room and equipment must be sterilised.

Mouse hepatitis virus (MHV) – a viral disease

The virus is present in many colonies showing little or no clinical effect. Occasionally liver lesions may be found at post mortem examination but serological testing is the only positive method of identification.

Enteric mouse hepatitis virus (EMHV) – a viral disease

This is a virulent virus causing violent diarrhoea and high mortality in infant mice. At necropsy the gut contains watery yellow faeces.

To ensure virus-free stocks virus free parents are maintained in virus proof isolators. Obtaining young by hysterectomy is not successful as infected females may pass the virus to their young via the placenta. Serological testing will demonstrate if the colony is virus free.

Mange – a mite infestation

Myocoptes musculinus – body mange, and *Myobia musculi* – head and neck mange, are the main ectoparasites of the mouse. Constant scratching, loss of hair and inflammation of the skin and in advanced cases the appearance of scabs are the usual signs. Mite have been successfully eradicated from mice using a tetraethylthiuriam monosulphide ('Tetmosol' ICI Ltd, Macclesfield, Cheshire) dip and repeating the treatment 14 days later or more recently by 'rolling' the mice in sulphur and repeating 14 days later. * The latter method is probably less stressful.

Nematodes – a roundworm infestation

Aspicularis tetraptera and *Syphacia obvelata* are commonly found in

* Green, C J & Needham, J R (1974) A simple technique for sulphur dressing mice. *Laboratory Animals* **8**, 327–328.

the mouse and may be seen macroscopically in the caecum. In small numbers they are not harmful but may increase as a result of overcrowding and poor husbandry causing diarrhoea and rectal prolapse. Piperazine citrate added to the drinking water and given for 1 week, withheld for 1 week and repeated for 1 week will reduce the infestation.

Diseases of the rat

Chronic respiratory disease (CRD or Endemic Pneumonia) – a viral disease

In the past it was thought that a number of bacteria and viruses were responsible for this disease; it is now believed to be a combination of two separate diseases: infectious catarrh caused by a mycoplasma and enzootic bronchiectasis caused by a virus. Infected animals can be heard to 'snuffle' and 'chatter'. Transmission is by direct contact with infected rats or mice when the organisms are introduced directly into the nasal passages, by the same means young rats are readily infected by their parents. At post mortem examination the lungs show lesions and may be consolidated, the nasal passages are inflamed. Inflammation of the middle ear may be present but this is usually due to a secondary invasion of *Streptobacillus moniliformis*. Culling obviously infected animals will not eradicate the disease and it is not treatable with drugs. CRD free animals can be bred from germ free or SPF parents and must be maintained behind a barrier away from conventional rat and mouse stocks.

Labyrinthitis – a bacterial disease

Apart from the causal agent *Streptobacillus moniliformis* it is believed that a mycoplasma is also involved. Affected animals tend to move in circles with the head tilted to one side. The disease is spread via the respiratory tract of infected animals. At post mortem examination there is inflammation of the upper respiratory tract and pus may be found in the middle ear. To control the disease affected or suspect animals should be culled.

Paratyphoid – a bacterial disease

The causal bacteria are *Salmonella enteritidis* and *S. typhimurium*. Symptoms are loss of condition, diarrhoea and death. At post mortem examination there is inflammation of the gut, enlarged spleen and haemorrhage of the lungs. The disease may be minimised by practising high standards of hygiene, care should be taken to ensure that food and bedding cannot be contaminated by wild rodents. Sick animals should be destroyed as they may recover to become carriers and reinfect stocks.

Nematodes – a roundworm infestation

Syphacia obvelata is commonly found in conventional rat colonies. At necropsy the worms can be seen macroscopically in the caecum. In small numbers they are not harmful but in quantity may cause rectal prolapse and diarrhoea in young rats. Piperazine acid citrate added to the drinking water and given for one week, withheld for one week and repeated for one week will reduce the infestation. The presence of *Syphacia obvelata* is often the cause of SPF colonies being downgraded.

Body mange – a mite infestation

The causal agents are *Myobia ratti*, *Myobia musculi* and *Notoedres* species. Affected animals scratch themselves giving rise to bald patches and scabs. A tetraethylthiuriam monosulphate ('Tetmosol' ICI Ltd, Macclesfield, Cheshire) dip used twice with a 14-day interval will eradicate the infestation.

Ringtail – a physiological disease

The cause of this condition is not clear but rats kept in an atmosphere of less than 50% relative humidity appear to be susceptible. The tail becomes ringed by a series of constrictions along its length and in extreme cases may slough off. Young rats are usually the first to be affected. (There is some evidence that the more humid microenvironment of a solid bottomed cage reduces the incidence.)

Diseases of the guinea-pig

Pseudotuberculosis – a bacterial disease

The causal agent is *Yersinia pseudotuberculosis* and the disease is usually contracted from green foods contaminated by droppings from infected birds. There is an acute and chronic form of the disease; acute cases die without warning usually as a result of rupture of mesenteric gland lesions into the mesenteric vein, chronic cases become emaciated and die in 3–4 weeks from the onset of the disease. At post-mortem examination nodular caseous lesions are seen in the mesenteric and colonic lymph nodes and viscera particularly the Peyer's patches of the small intestine. Lesions may also be seen on the lungs and pleura. To control the disease it is usual to kill all suspects and contacts and observe good hygiene routines including avoidance of fresh green food. Control may be achieved by vaccinating with a heat killed polyvalent vaccine.

Salmonellosis – a bacterial disease.

The causal agent is usually *Salmonella typhimurium.*

The main route of infection is oral by ingestion of contaminated material and the disease occurs as an acute or chronic condition: in acute cases a diffuse enteritis causing diarrhoea is usually followed by death. In chronic cases the animals become emaciated, death following after several weeks. At post mortem examination in acute cases an enlarged spleen and diffuse enteritis will be seen. The spleen may also be enlarged in chronic cases and pale foci may be present on the liver, the mesenteric lymph nodes may be enlarged and there is sometimes a peritonitis. As the disease is usually well-established by the time it is diagnosed, rigorous culling and complete cleaning and fumigation of rooms must be carried out.

Streptococcal pneumonia – a bacterial disease

The causal agent is usually *Streptococcus pyogenes* which is usually inhaled.

The infected animals lose condition, cough, sneeze and there may be a watery discharge from the eyes and nose, the urine may contain blood (haematuria). Pregnant females sometimes abort and some animals may die with no symptoms at all. At post mortem examination there may be inflammation of the peritoneum, pleura and pericardium and often a yellowish exudate sometimes confined to the genital region. Rigorous culling and optimum environmental conditions may help control the disease which may not be apparent until the animals are stressed; this is often highlighted by minor surgery when many animals may develop pneumonia and die.

Lymphadenitis (cervical adenitis) – a bacterial disease

The causal agent is *Streptobacillus moniliformis* which infects the neck and cheek region causing large abscesses. These often regress but some may become very large and eventually rupture. This condition can be controlled by avoiding materials such as thistles in hay that may cause neck or throat lesions.

Coccidiosis – a protozoal disease

The causal agent is the protozoan *Eimeria caviae* and is usually contracted by ingestion of contaminated food. The guinea-pigs become listless, lose their appetite (anorexia), have diarrhoea and coccidial oocysts are shed in the faeces. At post mortem examination, haemorrhagic patches will be seen on the intestinal wall and there will be general enteritis-like symptoms. Sulpha drugs, for example sulphamezathene or sulphaquinoxaline in the drinking water, are effective in controlling the disease. The guinea-pigs are best kept on wire floors thus reducing the chance of reinfestation. Oocysts in cages can be killed with a 1% ammonia solution.

Lice – insect ectoparasites

There are several species of lice which infest guinea-pigs, for example *Gyropus ovalis, Glirocol porcelli* and *Treponema jenningsi*. Signs of infestation may not be apparent until the guinea-pig is killed or shaved. Lice may be found over the entire body although the head and neck region is the commonest site. Regular cage cleaning and treatment of the guinea-pigs with benzene hexachloride (BHC) or malathion powder should control the condition.

Physiological disease – intrauterine haemorrhage

This is a rather obscure condition thought to be the result of dietary or liver metabolism imbalance although streptococcial infection has also been implicated. Pregnant animals rapidly become unthrifty and a bloody discharge will be observed from the vulva. At post-mortem examination a large gravid uterus will be found with one or both horns full of blood and the fetuses will be dead. Little can be done to prevent this condition except possibly to reduce feed intake to prevent the animal becoming too fat.

Vitamin C deficiency

Guinea-pigs are one of the few mammals that are unable to synthesise their own vitamin C and will therefore show signs of deficiency if dietary levels are inadequate. They become extremely emaciated, wounds do not heal, and there will be haemorrhage from eyes and gums followed by death. At post-mortem examination petechial haemorrhages will be seen on all serous surfaces and there will be a characteristic enlargement of the cortochondrial junction of the ribs. Adequate levels of vitamin C (as ascorbic acid) in the diet will prevent the condition i.e. 90–105 mg of ascorbic acid per 100 g of diet.

Diseases of the Syrian hamster

Salmonellosis – a bacterial disease

The organisms associated with salmonellosis are *Salmonella enteritidis* and *S. typhimurium*. An animal is most likely to become infected by ingesting organisms from the faeces of animal or human carriers.

In the acute stage the animal loses weight and condition rapidly and soft light-coloured faeces are seen; conjunctivitis may also be apparent. Death occurs between seven and fourteen days after infection. An animal with chronic symptoms may appear to recover although it will continue to excrete *Salmonella* organisms in the faeces for a long period. Once the disease is detected the colony should be sacrificed and rooms and equipment sterilised.

At post mortem examination the lungs may appear slightly grey in colour with some haemorrhaging and the liver shows areas of necrosis which may be multi-focal. The organism will be identified by bacteriological culture.

For treatment to be successful it must be administered very shortly after ingestion of the organism. Salmonellosis may be treated with antibiotics when there is evidence of invasion in the blood-stream although the effect is doubtful because some workers have found human patients treated with antibiotics may excrete the organism longer than untreated patients. Vaccination has proved effective in the immunisation of calves. Control measures such as the use of uncontaminated food and bedding, good hygiene, and the isolation of infected cases will prove helpful in reducing the incidence of the disease.

Wet tail – a bacterial disease

Conditions favourable to *Escherichia coli* and *Proteus vulgaris*, the organisms associated with Wet-tail, are improper housing, inadequate diet and poor husbandry. Symptoms of the disease are emaciation and weakness with the anal area becoming wet and discoloured.

At post mortem examination there is severe inflammation of the intestine which should not be confused with the presence of intestinal parasites such as *Hymenolepis nana*, which also cause inflammation.

To control the disease overcrowding should be avoided and the animals should be kept on sterilised dry bedding. If a good quality standard formula diet is fed, nutritional problems will be avoided.

Cestodes – a tapeworm infestation

Several internal parasites have been reported to cause disease in hamster colonies. The most common is *Hymenolepis nana*, known as the dwarf tapeworm and also found in rats and mice. It is responsible for causing intestinal obstructions and retarded growth in young animals. An intermediate host is the rat flea *Nosopsyllus fasciatus*. Treatment with a single oral dose of niclosamide ('Yomesan' Bayer UK Ltd Agrochem Division, Bury St Edmunds, Suffolk) at 200 mg/kg is effective against adult tapeworms only and it is advisable to repeat the procedure within 10–14 days.

Flagellates of *Giardia spp.* can cause digestive disturbance when present in large numbers in the small intestine, although their pathogenicity is low. Some hamster breeders have however, experienced several problems arising from *Giardia* infestation.

Body mange – a mite infestation

Two *Demodex* species of mite are commonly found in hamsters: *Demodex cricetis* burrowing into the epidermal pits and *D. aurati* into the hair follicles, causing few signs of disease other than mild hair loss (alopecia). These infestations can be treated by briefly immersing the hamsters in a 10% solution of tetraethylthiuriam monosulphide ('Tetmosol', ICI Ltd, Macclesfield, Cheshire).

Fighting injuries – a physical disease

Hamsters can be very aggressive and females will often attack and wound males. Hamster bites can penetrate deeply into the flesh and become infected if early treatment is not forthcoming.

Diseases of the rabbit

Infectious nasal catarrh (snuffles) – a bacterial disease

Two principal species of bacteria have been isolated in this condition, *Bordatella bronchiseptica* and *Pasteurella septica*, which invade the respiratory system usually from droplets produced by other sneezing rabbits.

The symptoms are characteristic and analogous to those found in the common cold in humans. The animal sneezes, it may breathe noisily, there is a muco-purulent discharge from the nasal passage, and it frequently wipes its nose with the inside of its fore-paws giving tell-tale wet patches on the paws. At post-mortem examination there may be mild inflammation of the nasal passages, severe rhinitis, sinusitis and pleuropneumonia.

The most effective method of control is to cull affected animals, sterilise cages and house all other animals singly to reduce spread of the infection. Limited success in preventing the disease has been achieved by a polyvalent vaccine.

Mucoid enteritis

This is an obscure condition of unknown cause but is sometimes a sequel to coccidiosis or stress particularly in early post weanlings. Death is preceded by mucoid diarrhoea and weakness of the hind limbs. Post-mortem examination reveals a typical enteritis with the lower gut containing variable amounts of mucoid faeces.

There is no effective cure for this condition but a coccidiostat, for example sulphamezathene in the drinking water or feed, is a sensible precaution providing this will not interfere with experimental programmes.

Syphilis – a bacterial disease

This is caused by *Treponema cuniculi* and is contracted during copulation. Ulcerative lesions of varying severity appear on the penis and prepuce of the male and on the vulva and vagina of the female. The anus, eyelids and mouth may also be affected. The condition can usually be cured by injecting neosalvasan or penicillin.

Coccidiosis – a protozoan disease

This is caused by the parasitic protozoans *Eimeria perforans* and *Eimeria magna* in the intestine and *Eimeria stediae* in the liver. It is contracted by ingesting oocysts from the faeces of infested rabbits. A rabbit suffering from intestinal coccidiosis may show a general loss of condition, some diarrhoea and finally death. Faecal examination will reveal oocysts. Post-mortem examination shows small infected foci on the wall of the intestine. The organisms can be seen in the gut wall when examined histologically. The hepatic form causes faulty metabolism which results in loss of weight and death. Either form can be eradicated by the following regime:

- treatment with a sulpha drug for example sulphamezathene which can be administered in the drinking water.
- sterilising cages or cleaning with a strong ammonia solution to kill oocysts.
- using wire floored cages to reduce the risk of reinfestation from the rabbit's own faeces.
- providing a coccidiostat in the diet or drinking water.

Myxomatosis – a viral disease

This is a disease that was first introduced to Australia when rabbits were deliberately infected to reduce the massive wild population. The disease eventually spread across Europe and finally reached the Kent coast in the early 1950s. It is spread by the mosquito and the rabbit flea. The symptoms are most unpleasant, the eyelids become congested and swollen followed by a conjunctivitis with a purulent discharge. The lips, nose and genitals may also become affected. There are few internal signs of the disease, the spleen may become enlarged and subcutaneous tumours may develop. Rabbits can be vaccinated against myxomatisis and if colonies are at risk this is advisable.

Ear canker – a mite infestation

The causal agents are *Chorioptes cuniculi* or *Psoroptes cuniculi*.
The rabbit shows signs of distress, irritation of the ear and if untreated the ear gradually becomes filled with crusty scabs from the exudative lesions caused by the mite. Regular inspection is important and as canker is easily controlled if the condition is observed. Affected animals can be treated with various insecticides such as a 1:3 aqueous solution of tetraethylthiuriam monosulphide ('Tetmosol', ICI Ltd) which can be applied to the affected area from a wash bottle. In very bad cases a second application a few days later may be necessary. Canker may also be responsible for secondary bacterial infections which may cause middle-ear disease.

Diseases of the ferret

Neck and face abscesses – a bacterial disease

Small wounds caused by agressive play may turn septic and produce an abscess, usually around the mouth or face. The abscesses may be drained surgically and the animal treated with antibiotics effective against the causal organism *Staphylococcus pyogenes*. In severe cases the facial bones may become affected by osteomyelitis and the animal should be killed as there is no effective treatment.

Tuberculosis – a bacterial disease

Three strains of *Mycobacterium tuberculosis*, human, avian and bovine, may affect ferrets. The disease is usually contracted from materials contaminated with bird droppings or from tuberculin-infected milk. The tuberculin testing (TT) scheme has almost eradicated the disease from cattle and therefore milk is rarely a source of infection. Care should be taken to isolate ferrets from contact with newly imported monkeys or any other possible sources of tuberculosis.

Symptoms are a rapid loss of weight and condition accompanied by a swelling of the mesenteric lymph glands. There is no treatment for the disease and affected animals should be killed, the bodies burnt and cages and equipment sterilised.

Canine distemper – a viral disease

The ferret's susceptibility to distemper led to its being chosen as the animal model in which to study the disease. The appearance of distemper in ferret colonies today is usually as a result of direct or indirect contact with dogs excreting the virus. The symptoms are discharges from the nose and eyes followed by muscular tremors and convulsions as the disease progresses. The only effective control is to kill all suspect animals, vaccination may give some protection to the others. All cages and equipment should be cleaned and sterilised.

Mange – a mite infestation

The mite *Otodectes cynotei* infests the ears and its presence is indicated by the animal scratching the irritation and causing inflammation. Benzyl benzoate may be effective in eliminating the mite.

A mange mite is also responsible for footrot causing swelling and the formation of scabs. The scabs should be soaked and carefully removed and benzyl benzoate applied to the affected areas. Bedding from the cages of infested animals should be carefully removed and burnt to prevent the mites being scattered about the room. The cages should be washed while still in the room with a solution of hot washing soda.

Diseases of the cat

Pneumonia – a bacterial disease

The causal agent of pneumonia is *Bordetella bronchiseptica* which is associated with the same disease in the dog, rabbit and guinea-pig. To avoid cross-infection the species should not be housed in close proximity in the animal unit. The signs of the disease are nasal discharge, dyspnoea and occasionally death. At post-mortem examination the lungs may be congested and show small areas of consolidation. The disease may be prevented by vaccinating kittens at six weeks of age followed by an annual booster injection. Some degree of control may be obtained by administering antibiotics, although vaccination is preferred as a long-term measure.

Septicaemia – a bacterial disease

The infection is caused by *Streptococcus haemolyticus* and the organism enters the body via the umbilical cord from the mother. The kittens are infected while they are being cleaned immediately after birth. The symptoms occur mainly in kittens although older cats may become infected from them. There is a rise in temperature, loss of appetite and associated respiratory distress. The disease may be controlled with a broad-spectrum antibiotic such as ampicillin ('Penbritin', Beecham, Brentford).

Feline panleucopaenia – infectious feline enteritis – a viral disease

This disease is caused by a small DNA virus which attacks all members of the cat family. It is a contagious disease resulting in sudden loss of appetite, a rise in body temperature, vomiting, diarrhoea and dehydration. The virus is spread from secretions and excretions of infected animals and is passed on through direct contact. At post-mortem examination signs of dehydration are evident and the lymph nodes may be enlarged and necrotic. Part of the ileum may appear reddened and occasionally the small intestine is involved.

Prevention of the disease is by vaccination – kittens are vaccinated at 6 weeks of age and this is followed by a booster injection two weeks later. Adult cats should be vaccinated annually.

Feline pneumonitis – a viral disease

The virus which causes pneumonitis is a member of the herpes group. It is disseminated over a large area by sneezing cats and therefore the route of infection is most probably respiratory. The cat begins to sneeze frequently, and this is followed by the appearance of conjunctivitis and fever, which may cause the temperature to rise to 40.6°C (105°F). Discharges from the nose and eyes become obvious and as the disease progresses the animal loses its appetite and becomes depressed. To help control the disease broad-spectrum antibiotics are effective against secondary bacterial infections.

Nematodes – a roundworm infestation

The common roundworms that infest cats are *Toxascaris leonina* and *Toxascaris mystax*. The ova are ingested by the cat and the roundworms develop in the intestine where they continue to flourish in ideal conditions. The symptoms are not always obvious, but when heavy infestations occur the animal may vomit and lose weight and condition. An infestation may be controlled by piperazine citrate administered orally and the treatment should be repeated to ensure complete eradication. Diagnosis can be made by microscopic examination of the faeces for adult worms and ova.

Cestodes – a tapeworm infestation

Common species found in the cat are *Taenia taeniaeformis*, *Taenia pisiformis* and *Dipylidium caninum*; the cat is infested by eating the tapeworm cyst found in the flesh of mice and rats, which act as the intermediate hosts for the tapeworm. The cat flea (*Ctenocephalides felis*) is the intermediate host for *Dipylidium caninum* – the cat becomes infested by eating the fleas when grooming. A variety of anthelmintics are available, the most useful being arecoline hydrobromide and arecoline acetarsol. The manufacturers' instructions should be followed carefully when administering treatment for complete eradication of cestodes.

Body mange – a mite infestation

Notoedres cati and *Demodex folliculorum* are the causal agents of body mange and they are contracted from the hair of other infested cats. The symptoms of body mange are hair loss and small crusty lesions which appear on the face and ears and caused by the mite burrowing under the skin. Infested animals should be removed from the unit and dusted with malathion powder. Never use benzyl benzoate; it is contra-indicated in cats because it seriously affects the central nervous system.

Ear canker – a mite infestation

The organisms *Notoedres cati* and *Demodex folliculorum* which cause body mange are also responsible for ear canker. Other infected cats provide the source of contamination. The cat exhibits symptoms of obvious irritation by scratching its ears and shaking its head. On inspection the ears contain a dark brown waxy mass of mite eggs and larval debris, which can be treated by dibutyl phthalate dropped into the ear.

Ringworm – a fungal disease

The causal agents of ringworm in the cat are *Trichophyton felineum* and *Microsporum canis*. The route of infection is by direct contact with infected animals. Skin lesions appear on various parts of the body, particularly on the forepaws, face and neck. The typical ringworm lesion is a circular formation of crusty scabs. Administration of griseofulvin tablets will eradicate the infection over a period of weeks. Infected or suspect animals must be isolated from the colony before treatment begins. Ringworm is a zoonotic disease.

Urinary calculi – a nutritional disease

The urethra and neck of the bladder in male cats become obstructed with large hard deposits of minerals, calculi or bladder stones. Affected cats have difficulty urinating, passing only a small volume of urine which is often blood-stained. The calculi are composed mainly of calcium and require surgical removal when they are detected. Many remedies have been tried unsuccessfully until a diet containing 1% sodium chloride was fed, which resulted in a reduced incidence of the number of cats with calculi.

Diseases of the dog

Distemper – a viral disease

Distemper and hard pad are caused by the same virus and can cause serious losses in colonies housed in the same building. Signs of distemper are loss of appetite, fever, and discharge from the eyes and nose and it is this discharge which can infect other dogs. Diarrhoea and respiratory distress follow the initial symptoms, and fits and convulsions may also occur. Distemper can be prevented by vaccinating puppies at 9–12 weeks of age, with annual booster injections for continued protection.

Hard pad – a viral disease

The infection becomes obvious from the swelling and thickening of the pads which tend to make a tapping sound when the dog walks. Other symptoms are similar to those described for distemper.

Infectious hepatitis – a viral disease

The virus, which affects the liver, is transmitted through the saliva or urine of an infected dog and a sick animal will display loss of appetite and appear depressed, the water intake may rise considerably and the body temperature may rise to 40°C (104°F). Prevention of the disease is by vaccination, usually given in conjunction with distemper and hard pad vaccines.

Leptospirosis – (jaundice) – a spirochaetal disease

Leptospirosis is a zoonotic disease caused by *Leptospira canicola* and *Leptospira icterohaemorrhagiae*, two spirochaetes which are more closely related to bacteria than to other organisms. The animal becomes infected from the urine of infected rats or from eating the carcase; symptoms are loss of appetite, muscular stiffness and pain, diarrhoea and conjunctivitis followed by death. The disease may be prevented by vaccination of puppies followed by annual booster injections in adult dogs.

Parvo virus infection

This disease, caused by a corno-virus, was first reported in the USA in 1978 and since then has spread throughout the world. Dogs of all ages may be affected but the severest attacks are seen in puppies between 4–11 weeks.

Infected dogs show signs of anorexia, depression and vomiting and after a few days lose weight dramatically and produce a bloody diarrhoea. A yellow discharge from the eyes may also be seen.

At post mortem examination severe breakdown of the lining of the small gut may be seen with consequent rupture of small blood vessels. Oedema of the heart and kidney, myocarditis and sometimes lesions of the lung and liver may also be observed.

Treatment with guanamycin, neomycin, penicillin and fluid replacement therapy along with careful feeding with milk, creamed or baby rice etc. has been successful. Vaccination with inactivated feline panleucopaenia vaccine has also been successful in controlling outbreaks of the disease.

Rabies – a viral disease

Rabies is an acute, fatal and zoonotic disease caused by a neurotropic virus. It is transmitted by the bite of a rabid animal or by the saliva of an infected animal entering the body for instance through a cut. The symptoms of the disease appear in two phases: the first is when the normal behaviour of the dog changes perhaps from the affectionate to unaffectionate and may become hypersensitive. The second stage may be in the form of paralysis where the animal cannot keep its jaws closed, the pupils may constrict causing the dog's expression to change. Sometimes the dog may roam aimlessly and try to bite any object that it encounters.

Dogs suspected of rabies must be confined and help sought from the Ministry of Agriculture, Fisheries and Food Divisional Veterinary Office. Vaccines and sera are available for treatment of rabies but success is not guaranteed. The *Rabies (Importation of Dogs, Cats and other Mammals) Order 1974* is part of the policy to prevent the introduction of rabies into this country. The law states that amongst other classified species dogs and cats are subject to a quarantine period of six months.

Nematodes – a roundworm infestation

The causal agents are *Toxacara canis* and *Toxascaris leonina*. It is not always obvious to the untrained eye that these agents are present because dogs may not exhibit symptoms of an infestation. Puppies sometimes become 'pot-bellied', and if the condition is not treated they lose weight and condition and may even die. Adult worms are commonly excreted in the faeces and the dog may show its irritation by dragging its anus along the ground in an effort to rid itself of worms. The dog becomes infected by faecal contamination from other dogs. Roundworm infestations can be controlled by administering piperazine acid citrate. Bitches should be wormed during gestation to prevent puppies being infected at birth. Puppies should be wormed at weaning.

Cestodes – a tapeworm infestation

The infestation may be caused by a number of different tapeworms – *Taenia pisiformis*, *Taenia hydatigena*, *Dipylidium caninum*. The route of infection for *Taenia* spp. involves an intermediate host, the rabbit flea. The dog eats the cyst in the infested rabbit and the worms develop in the intestine. Heavy infestations may cause loss of weight and condition and even death. Control is effected by eradicating fleas and by using cooked or tinned food instead of raw meats. Rodents as intermediate hosts of *Dipylidium* are a source of infestation and should be prevented from entering the kennels. Repeated treatment with anthelmintics will keep dogs healthy and relatively worm free.

Ear canker – a mite infestation

Fleas, ticks and lice will live on dogs in the first instance without causing too much discomfort. Ear canker is caused by the mites *Sarcoptes scabiei* or *Otodectes cynotis*. The animal may shake its head and scratch its ears, often making them bleed, in sheer irritation. On inspection the ears may show a dark mass of wax and debris from dead mites. Canker is treated by cleaning the ears and applying an acaricide, for example dibutyl phthalate.

Body mange – a mite infestation

Body mange is a contagious disease spread by direct contact with infested animals or equipment that has been used for grooming. A common cause of sarcoptic mange is the mite *Sarcoptes scabiei*. Demodectic mange is caused by *Demodex canis*. The animal scratches persistently at the infested areas and the skin becomes dry and wrinkled. Often crusts and lesions are obvious around the head, eyes and ears, but severe infestation can lead to the mites spreading to the neck and abdomen.

To treat mange infestations the dog should be clipped to expose the lesions and bathed in an acaricide such as benzyl benzoate or benzene hexachloride. It is advisable to provide a regular bathing programme for dogs to prevent mange and it is particularly important to bath and treat fresh importations of stock animals.

Diseases of simians

The diseases described are zoonotic and the precautions described in Chapter 19 should be taken by the animal technician to avoid contracting disease.

Dysentery – a bacterial disease

The causal organisms *Shigella flexner*; *Sh.sonnei* and *Sh.schmitzei*. The disease is acquired from other monkeys or occasionally man. The incidence is high in newly-imported monkeys as a result of contact with diseased animals in overcrowded holding areas where they are captured and the stress of the journey to this country. The classic diagnostic feature is the presence of mucus and blood in the fluid stools. The animals become dehydrated and emaciated and often die.

The infection may be controlled by the administration of antibiotics, single housing and strict hygiene.

Tuberculosis – a bacterial disease

The causal organisms are *Mycobacterium tuberculosis* of the human and bovine types. Newly imported monkeys should be screened three times for the disease while in quarantine. The standard methods are by X-ray and the use of the Mantoux Test. In the Mantoux Test a small amount of diluted tuberculin antigen is injected into the eyelid of the monkey and if after three days the eyelid becomes reddened and swollen, the reaction is positive. There may be false positives and false negatives, hence the necessity of repeating the test twice more. *Any* repeatedly positive reactor should be destroyed as there is no practical way of treating infected animals and the risk to staff is too high to be tolerated.

On commencing work with monkeys and regularly thereafter, technicians should be X-rayed to detect and prevent cross-infection.

Pneumonia – a bacterial disease

The causal organisms are *Diplococcus pneumonia* and *Klebsiella pneumoniae*. This disease may also be contracted from infected man and from other monkeys. Affected animals exhibit repiratory distress and general loss of condition. The disease may be diagnosed by X-ray examination or at post mortem by bacterial examination of the respiratory tract.

Rabies – a viral disease

The causal organism is a neurotropic virus. In recent years this disease has received wide publicity and presents a hazard equal to that of B virus. All monkeys are potential carriers of rabies and strict regard must be paid to the quarantine conditions under which imported monkeys are housed. Minimum standards are mandatory under the *Import of Mammals Order 1974*. If humans contract rabies there is little chance of recovery. However, vaccines are available and only staff so protected should be permitted to work with monkeys.

B virus – a viral disease

The causal organism is *Herpesvirus simiae*. The virus is related to the 'cold sore' virus affecting humans (*Herpesvirus simplex*) which is known to be fatal to marmosets. B virus affected monkeys exhibit sore lesions on the lips, inside the mouth and on the tongue, which become ulcerated and form scabs. After a few weeks the scabs disappear and the animal appears normal once more but remains a carrier. In humans the disease is usually fatal and may be contracted through bites or scratches or even by handling of contaminated equipment.

Marburg virus (Green Monkey Disease) – *probably a viral disease*

The causal organism is thought to be a virus but it has not been fully identified. The disease is named after an outbreak in Marburg, Germany in which several people died as a result of coming into indirect contact with a batch of newly-imported monkeys. The same precautions should be taken as described in Chapter 19.

Diseases of chickens

Salmonellosis – a bacterial disease.

The causal agents are either the highly host-adapted *Salmonella* species, e.g. *S.pullorum S.gallinarium*, or by non-host adapted species such as *S.typhimurium*.

Pullorum disease: This is caused by the bacteria *Salmonella pullorum* and can cause high mortality among young chicks. Transmission is mainly via the egg but may also occur from contact with infected faeces. Affected birds huddle near a heat source, do not eat, appear listless and whitish faeces may be seen around the vent. At post-mortem examination there is focal necrosis of the liver and spleen, and grey nodules on the lungs and heart and gizzard muscles. Adult carriers may not show outward clinical symptoms but usually have peritonitis or pericarditis.

Fowl typhoid: Caused by *Salmonella gallinarium* and like *S.pullorum* may be transmitted in the egg and produces similar effects in young birds. It has a greater tendency to be transmitted among older birds; mortality at all ages is high. At post-mortem examination there may be a swollen, friable liver with or without necrotic foci, enlarged spleen and kidneys, anaemia and enteritis.

The same treatment is recommended for these two diseases. Various antibiotics and sulphanomides will reduce mortality; furazolidone is the most effective medication. Rigorous cage hygiene and severe culling of affected birds are also recommended.

Chronic respiratory disease (CRD) – a bacterial disease

A number of organisms have been implicated with CRD, including *Escheria coli, Haemophilus gallinarum* and *Mycoplasma gallisepticum*. The symptoms vary according to the organism involved but some degree of respiratory distress will be seen in all cases. Antibiotics have been used with success in controlling some outbreaks of CRD but, under normal conditions, it is better to kill affected and in-contact birds and to clean and sterilise the room and cages.

Infectious laryngotracheitis (ILT) – a viral disease

The causal agent is a virus. This is a highly contagious disease; affected birds will cough, sneeze, gasp, rattle, and may extend their necks during inspiration. At post-mortem examination bloody mucus and a yellow, caseous exudate will be seen in the trachea. Vaccines are available; strains of low virulence are given in eye drops, whereas the virulent strains are applied by brush to the vent mucosa. In all cases vaccination is normally carried out between 5 and 16 weeks of age.

Newcastle disease (fowl pest) – a viral disease

The causal agent is Newcastle disease virus (NDV). The disease occurs in two forms, mild and peracute. The mild form particularly affects young chicks which show a number of symptoms; respiratory distress, coughing, loss of appetite, nervous symptoms and even complete paralysis. Egg production is considerably reduced in laying birds and on recovery thin-shelled and misshapen eggs are produced. In the peracute form of the disease, death may occur before symptoms are observed. However, marked respiratory symptoms may be seen along with a general dullness, loss of appetite, increased thirst and a profuse yellow-green diarrhoea. There are often no obvious lesions to be seen at post-mortem examination.

The disease is notifiable under the *Disease of Animals Act 1950* but it is no longer compulsory to slaughter infected and contact birds.

Vaccines of inactivated viruses are now widely used to control the disease.

Fowl pox – a viral disease

This disease is also known as avian diphtheria. There are two forms of this disease, one producing crusty pox lesions on the comb and wattles, and the other patchy, yellow lesions inside the mouth. Severe outbreaks may result in a high mortality. Local treatment of the lesions may be of some value but prevention by vaccination at about three months of age is preferable.

Avian leukosis complex (including Marek's disease) – a viral disease

Avian leukosis is a complex of three groups of neoplastic disease. Within this complex the two diseases of major economic importance are Marek's disease and lymphoid leukosis.

Marek's disease (fowl paralysis): A viral disease caused by a DNA virus of the B group of herpes virus. Marek's disease is highly infectious being transmitted on feather dander from infected chickens. Most conventional flocks carry the infection but develop resistance to it. The occurrence of the clinical disease depends on the age and genetically-determined resistance of the birds, the dose of virus received and the virulence of the virus strain.

The disease usually occurs at 12–14 weeks of age. Birds become lame and paralysed; wings, neck and eyelids may also be affected. In younger birds anorexia and depression will be seen along with neoplastic lesions on the skin and muscle. At post-mortem examination nerves may be enlarged and lymphomas occur on many of the organs. The disease is fatal. A live vaccine has been developed to protect birds from this disease and chicks are usually vaccinated when one day old.

Lymphoid leukosis: Lymphoid leukosis is caused by an RNA virus which induces the neoplastic transformation of lymphocytes within the follicles of the bursa of Fabricius. The virus is transmitted via faeces and saliva of infected birds. Chicks are most susceptible during the hatching and brooding periods, although chicks may gain temporary passive immunity from maternal antibodies. Obvious symptoms may not be apparent until the chick is found dead. At post-mortem examination small tumours of the bursa of Fabricius may be seen and in advanced cases massive enlargement of the whole organ. Metastases may occur in many other organs. No control measures other than strict hygiene and elimination of infected birds and contacts have been successful.

Avian encephalomyelitis (Epidemic tremors) – a viral disease

This is caused by an enterovirus and is usually seen in young chicks from 1–6 weeks of age. The first symptoms are dull eyes, difficulty in walking and the affected chick will eventually spend most of its time resting on its haunches. Nervous tremors of the head and neck may be seen. Muscle incoordination becomes progressively worse and death usually follows. In older birds the only sign of the disease may be a drop in egg production.

Chicks are usually infected with the virus in the egg from carrier mothers. These carriers may develop a degree of immunity which may be passed on to the chick in the egg. In this case, the chick would be protected for its first few weeks of life. If the chick becomes infected later it will probably be an asymptomatic carrier.

Vaccines against the disease are available and these give lasting immunity.

Coccidiosis – a protozoan disease

The disease may be caused by any one of a number of the protozoan *Eimeria* species. The organisms invade the caeca or the intestine.

Caecal coccidiosis caused by *Eimeria tenella* is usually found in young chicks. Blood-stained faeces are produced and at post-mortem examination the caeca will be found to contain blood and blood clots. Mortality may be as high as 50%. Most coccidiostats will be effective against this species.

Intestinal coccidiosis is usually associated with *Eimera necatrix* but many other species also affect the intestine. Chickens become infected by ingesting contaminated faeces. The birds will generally be in poor condition, will not eat and will show signs of diarrhoea. Mortality is variable according to the level of infection and the species of organism responsible. At post-mortem examination, haemorrhagic lesions are found at the site of the infection.

Oocysts may sporulate in warm, humid conditions and therefore rigorous cleaning and dry bedding will reduce the risk of infection. Oocysts can be killed by washing out clean cages or pens with a 1% ammonia solution. Various drugs are available that are effective against coccidia and may be added at prophylactic levels to feed or drinking water.

Nematodes – a roundworm infestation

Many species of nematode infest poulty: perhaps the commonest and most important are the gapeworm *Syngamus trachea*, and the large roundworm, *Ascaridia lineata*.

The gapeworm *Syngamus trachea* invades the respiratory system, particularly the trachea; infected birds show signs of respiratory distress and worms are sometimes coughed up. Heavy infestations are usually fatal. Apart from ingesting infective ova from contaminated surroundings, chickens may also ingest ova from infected earthworms, slugs and snails, vectors that would not be important in a cage system. Proprietary anthelmintics are available against gapeworms and strict pen hygiene will help eliminate this parasite.

Ascarid worms are relatively common in conventional poultry. Heavy infestation will cause poor growth, poor egg production and sometimes death. At post-mortem examination worms will be found in the intestine where haemorrhage and enteritis may be seen. As the eggs of ascarids can survive for several months on cool, damp ground, dry bedding is important for birds in floor pens. Proprietary anthelmintics are available for treatment of infected birds; most contain piperazine compounds.

Cestodes – a tapeworm infestation

Several species of cestode are found in the intestine of chicks, the commonest being the *Raillietina* and *Hymenolepus* spp. As with most endoparasites, infected birds will appear unthrifty and worm segments may be found in the faeces. The voided segments are not infective to poultry but contain eggs which will develop into infective larvae once inside an intermediate host such as an earthworm or snail.

At post-mortem examination the worms will be seen attached to the wall of the intestine with associated signs of haemorrhage and enteritis. Once again, good cage hygiene will be effective in reducing the risk of infestation, and birds can be treated with proprietary anthelmintics, usually administered in the drinking water.

Lice

Lice are the commonest ectoparasites of poultry: there are many species of biting lice but no sucking lice that infest poultry. The commonest are the body louse *Menacanthus stramineus* on the skin, the shaft louse *Menopon gallinae* on feather shafts, the head louse *Cyclotogaster heterographus*, and the wing louse *Lipeurus caponis*. The life cycle of a louse is usually completed on the host, although man can act as a temporary carrier. Affected birds often show poor egg production or slow weight gain due to the irritation produced by the lice. Examination of the bird will reveal eggs or adult lice on the feathers under the wings or around the vent. Various insecticidal powders and sprays are available for controlling infestations.

Mite

Several species of mite infest poultry: the most common is the red mite, *Dermanyssus gallinae*. It is less than 1 mm long and grey in colour until it is engorged with the host's blood and becomes red. These mite do not live on the host but in cracks and crevices, coming out to feed on the chicken at night. They multiply very quickly in warm environments. Infestations of red mite will cause loss of condition and anaemia, particularly in young birds, and they may transmit other diseases such as fowl pox. Good cage hygiene and treatment of birds with an insecticide such as malathion will control mite infestations.

Calculations 22

Mathematics

A basic understanding of mathematics is desirable in any job but is essential for the animal unit technician. Some type of calculation is required every day whether it is a simple case of determining the percentage of animals born which are weaned, making up a solution of detergent or disinfectant, or something more complex such as calculating the amount of fumigant required for a particular room.

Percentage

Percentage is a description of part of a whole where the whole is compared to 100. By reducing various proportions to percentages a standard measure for comparison is given. For instance, two rooms hold a number of mice, some of which are found dead. It is reported that Room A held 80 mice but 20 are now dead and Room B held 900 mice but 162 are now dead. Obviously more mice died in Room B than in Room A, but in which room is it the more serious problem?

As a percentage of the totals held in the rooms, it can be seen at once that the death rate is higher in Room A.

Room A – 80 mice were held – 20 now dead.

$$80 \text{ mice} = 100\%$$

$$\therefore 1 \text{ mouse} = \frac{1}{80} \text{ (one eightieth of the total)}$$

$$\therefore 20 \text{ mice} = \frac{20}{80}$$

To express $\frac{20}{80}$ as a percentage multiply by $\frac{100}{1}$ *(note* $\frac{100}{1} = 100$).

$$\% \text{ dead} = \frac{20}{\cancel{80}_{20}} \times \frac{\cancel{100}^{25}}{1} \text{ } (\div \text{ by } 4)$$

$$\% \text{ dead} = \frac{\cancel{20}^{1}}{\cancel{20}_{1}} \times \frac{25}{1} \ (\div \text{ by } 20)$$

$$= \frac{1}{1} \times \frac{25}{1}$$

$$= 25\%$$

In room A 25% of the mice had died.

Room B – 900 mice were held – 162 now dead.

$$900 \text{ mice} = 100\%$$

$$\therefore 1 \text{ mouse} = \frac{1}{900} \text{ (one nine hundredth of the total)}$$

$$\therefore 162 \text{ mice} = \frac{162}{900}$$

To express $\frac{162}{900}$ as a percentage multiply by $\frac{100}{1}$ (note $\frac{100}{1} = 100$).

$$\% \text{ dead} = \frac{162}{\cancel{900}_{9}} \times \frac{\cancel{100}^{1}}{1} \ (\div \text{ by } 100)$$

$$= \frac{\cancel{162}^{18}}{\cancel{9}_{1}} \times \frac{1}{1} \ (\div \text{ by } 9)$$

$$= \frac{18}{1}$$

$$= 18\%$$

In room B 18% of the mice had died.

So in Room A 25% of the mice are dead and in Room B only 18%.

Example

In a colony of rats 720 are born and of these 600 are weaned. What is the preweaning loss expressed as a percentage?

$$\text{Number born} = 720$$
$$\text{Number weaned} = 600$$

$$\therefore \text{Preweaning loss} = 720–600 = 120$$

$$\% \text{ Preweaning loss} = \frac{120}{72\cancel{0}} \times \frac{10\cancel{0}}{1} \ (\div \text{ by } 10)$$

$$= \frac{\cancel{120}^{10}}{\cancel{72}_{6}} \times \frac{10}{1} \ (\div \text{ by } 12)$$

$$= \frac{10}{6} \times \frac{10}{1} (\times\ 10 \text{ by } 10)$$

$$= \frac{100}{6} (\div \text{ by } 6)$$

$$= 16.66\% \quad \text{Assume } 17\%.$$

Given a percentage one may be asked to find the value which it represents.

What is 12% of 250?

$$250 = 100\%$$

$$\therefore 1\% = \frac{250}{100}$$

$$\therefore 12\% = \frac{\cancel{250}^{10}}{\cancel{100}_{4}} \times \frac{12}{1} (\div \text{ by } 25)$$

$$= \frac{10}{4} \times \frac{12}{1} (\times\ 10 \text{ by } 12)$$

$$= \frac{120}{4} (\div\ 120 \text{ by } 4) = \frac{30}{1}$$

$$= 30$$

12% of 250 = 30

Example

In a certain strain of mice 14% are expected to abort before term. If 350 mice are pregnant how many can be expected to abort?

$$1\% = \frac{350}{100}$$

$$14\% = \frac{35\cancel{0}}{10\cancel{0}} \quad \frac{14}{1} (\div \text{ by } 10)$$

$$= \frac{35}{\cancel{10}_{5}} \times \frac{\cancel{14}^{7}}{1} (\div \text{ by } 2)$$

$$= \frac{\cancel{35}^{7}}{\cancel{5}_{1}} \times \frac{7}{1} (\div \text{ by } 5)$$

$$= \frac{7}{1} \times \frac{7}{1} (\times\ 7 \text{ by } 7)$$

$$= 49$$

49 mice may abort.

Given a number which is also a percentage, one may be asked to find the number which is equal to 100%.

If 80 is equal to 25% what is 100%?

$$25\% = 80$$

$$1\% = \frac{80}{25}$$

$$100\% = \frac{80}{\cancel{25}_1} \times \frac{\cancel{100}^4}{1} \ (\div \text{ by } 25)$$

$$= \frac{80}{1} \times \frac{4}{1} \ (\times\ 80 \text{ by } 4)$$

$$= 320$$

$$100\% = 320.$$

Example

60% of the hamsters on test must be fed experimental diet *A*; there is enough diet *A* for *54* hamsters, what is the total number on test?

$$60\% = 54$$

$$1\% = \frac{54}{60}$$

$$100\% = \frac{54}{6\cancel{0}} \times \frac{10\cancel{0}}{1} \ (\div \text{ by } 10)$$

$$= \frac{\cancel{54}^9}{\cancel{6}_1} \times \frac{10}{1} \ (\div \text{ by } 6)$$

$$= \frac{9}{1} \times \frac{10}{1} \ (\times\ 9 \text{ by } 10)$$

$$= \frac{90}{1}$$

Total on test = 90.

Average (arithmetical mean)

Average is the quantities added together and their sum (or total) divided by their number. To find the average of the following numbers:

4 7 5 9 3 2

Add: $4 + 7 + 5 + 9 + 3 + 2 = 30$ (sum or total)

The number of quantities is 6

$$\therefore \text{Average} = \frac{30}{6} (\div \text{ by } 6)$$

$$= \frac{5}{1}$$

Average = 5.

Example 1

5 mice had litters of 5, 6, 7, 8 and 9.

What is the average litter size?

Add: 5 + 6 + 7 + 8 + 9 = 35 (sum or total)

5 mice had 35 young

$$\therefore \text{Average for 1 mouse} = \frac{35}{5} (\div \text{ by } 5)$$

$$= \frac{7}{1}$$

The average litter size is 7.

Example 2

The number of rabbits on experiment at the end of each week over a 6-week period is 15, 12, 8, 17, 0 and 14.

What is the average number of rabbits on experiment each week?

15 + 12 + 8 + 17 + 0 + 14 = 66

$$\frac{66}{6} = 11$$

Average number of rabbits on experiment is 11.

Note: Although there were no animals on experiment one week the sum total must be divided by 6 (the total number of weeks).

Ratio

Ratio is the proportion of 2 quantities to one another, the number of times (whole or fractional) which one quantity is contained in another of the same kind. The ratio of 4 objects to 8 objects of the same kind is expressed 4:8 (4 is to 8) usually simplified by division to 1:2. Ratio is generally expressed like this in the lowest terms by dividing both terms by their highest common multiple (HCM), the largest number that will divide into both of them.

To express 4:8 in the lowest terms:

4 and 8 are divisible by 2 and 4, but 4 is the highest common multiple

(larger than 2). So 4 and 8 are divided by 4 to give the ratio in the lowest terms:

4:8 (÷ by HCM 4)
1:2

Example 1

In a litter of rabbits there are 3♂ and 6 ♀. What is the sex ratio of males to females?

Sex ratio ♂ : ♀ = 3:6 (÷ by HCM 3)
= 1:2
Sex ratio ♂: ♀ = 1:2

Example 2

A diet contains 250 kg of wheat, 200 kg of oats and 100 kg of skimmed milk powder. In what proportion is each ingredient present in the diet?

250:200:100 (÷ by HCM 50)
5:4:2

The proportion of wheat: oats: skimmed milk powder = 5:4:2

If given the proportions of ingredients for a diet and asked to make a certain amount one could determine the quantity of each ingredient required.

Example 3

Make up 50 kg of diet with the ingredients in the following proportions:
Wheat 14: Grassmeal 5: Oats 4: Fishmeal 2

Find the total of the number of proportions or parts.

$14 + 5 + 4 + 2 = 25$

Divide the amount of diet required by the number of parts to find the value of 1 part:

$$\frac{50}{25} = 2$$

The value of 1 part is 2 kg

Multiply the proportions by 2 to give the total amount of each ingredient required in kg:

Wheat	$14 \times 2 =$	28 kg
Grassmeal	$5 \times 2 =$	10 kg
Oats	$4 \times 2 =$	8 kg
Fishmeal	$2 \times 2 =$	4 kg
Total		50 kg

Breeding calculations

When animals are bred for experimental purposes a breeding programme must be devised. The programme is necessary not only to indicate how many animals must be mated to produce the number of young required but also to act as a guide for planning the amount of space the cages will occupy, the labour required and the amount of food to be ordered. Established programmes need revising when requirements change and new systems devised when different species or strains are required. The principles involved are similar when making calculations for either a practical or theoretical purpose. Usually the request is for a supply of a given number of animals at a particular age or weight at specific intervals. Before the calculations can be attempted the following facts must be known about the particular species required.

Length of oestrous cycle
Age at first mating
Gestation period
Average litter size
Age at weaning
Length of interval from parturition to remating
Economic breeding life

The economic breeding life of a female is the period of time when she is producing litters of good quality and quantity at regular intervals. This may be expressed in time such as 2 years for a rabbit, or in numbers of litters produced such as 6 litters for a rat or mouse. The economic breeding life of the male is determined by its ability to regularly achieve fertile matings that result in the expected litter size for the species. Animals that do not attain or subsequently fall short of these requirements are culled from the breeding colony. Seasonal breeders such as the ferret and some farm animals, and species with 1 or 2 heat periods per year such as the dog, do not fulfil these conditions, the useful breeding life being determined by the production of the individual.

Considerable thought should be given to the breeding system, for example whether the animals are placed in pairs, trios or harems, or if advantage will be taken of the post-partum oestrus. Economics will play an important part in the decision in that more space and labour are required if animals are housed in pairs as opposed to trios or harems. However, with species or strains that are difficult to breed the requirements of these animals will come first with space and labour playing less important roles. The harem system, using one male to a number of females, is economical of space and labour and is effective when producing random bred animals, for example guinea-pigs, when full advantage is taken of the post-partum oestrus. Rats and mice may be harem mated with pregnant animals removed to littering cages before parturition. Although the post-partum oestrus mating is missed the system can prove economical if single sex young are required. The unwanted sex can be killed at birth, and litters made up to an optimum number by fostering and surplus lactating females returned to the harem for remating.

When hybrids, crosses between parents of different strains, are required it is important to remember that pure line replacement breeders must also be maintained, for example if AB hybrids are required the programme must include A × A and B × B matings to produce pure line offspring for future breeders; the pure line cannot be reconstituted by mating hybrids together.

In practical terms an important consideration is the age or weight at which the animals are to be issued from the breeding unit. Considerably more space, labour and food are required if animals are to be issued at 3 or 4 months of age than if they are issued at weaning.

It is advisable to collate all available information about a strain or species not previously housed in the unit, before planning the breeding programme. For example whereas mice and rats will commonly have litters of 8–12, some inbred or special strains will breed reluctantly, producing litters of 2–3 at irregular intervals.

Finally when making the calculation round off figures so that they are simple to work with. Assume 50 weeks to 1 year and 4 weeks to 1 month. Where the animal data are concerned and ranges are involved use a figure to the advantage of the programme, for example if the litter range is 6–9 use 7 as the litter size, if the litter interval is 3.5 weeks use 4 weeks, or the number of young produced per female per week is 1.8 use 2. Thus a margin will be allowed for errors. If all goes well there is a possibility that more animals will be produced than required, this will be obvious early in the programme when adjustments can be made to reduce the output. It is better to have a small surplus of animals for a short period than waste valuable research time waiting for the required number.

Basic breeding calculations

To produce 40 guinea-pigs each week

Relevant data	Range	Figure for calculation
Length of oestrous cycle	14–16 days	16 days
Length of gestation	59–72 days	65 days
Litter size	3–6	4

Breeding system – harem system taking advantage of the post-partum oestrus groups of 1 male with 8 females

Litter interval = length of oestrous cycle + gestation in days
= 16 + 65 days
= 81 days

$$\text{Number of litters per } \text{♀} \text{ per year} = \frac{\text{number of days in a year}}{\text{litter interval in days}}$$

$$= \frac{365}{81}$$

= 4.5 (assume 4)

Number of young per ♀ per year = number of litters per year × number in litter

= 4 × 4

= 16

$$\text{Number of young per } \text{♀} \text{ per week} = \frac{\text{number of young per year}}{\text{number of weeks in year}}$$

$$= \frac{16}{50}$$

= 0.32 (assume 0.3)

Number of young required per week = 40

$$\therefore \text{ number of breeding females} = \frac{\text{number required per week}}{\text{number of young per } \text{♀} \text{ per week}}$$

$$= \frac{40}{0.3}$$

= 133.3 (assume 134)

The economic breeding life of a guinea-pig is 2 years. Therefore 1/24 of the breeders need to be replaced each month

$$\frac{1}{24} \text{ of } 134 = \frac{134}{24}$$

= 5.5 (assume 6)

To provide 6 extra young per month – assuming a sex ratio of 1♀:1♂ – 12 extra young will be required (3 per week)

Number of young per ♀ per week = 0.3
Therefore 10 females will supply 3 young per week
Total number of females required = 134 + 10
= 144

Fewer males are required as replacement male breeders and these could be taken from stock.

To produce 400 male rats per week

Relevant data	Range	Figure for calculation
Length of oestrous cycle	4–5 days	5 days
Length of gestation	21–22 days	22 days
Litter size	8–10	8
Weaning age	21 days	21 days

Breeding system – harem system in groups of 1 male with 6 females, removing the females prior to parturition, making up litters of 8 male pups and fostering to one female, returning half the breeding females to the harem cage for remating (female young except those required as replacements for breeding females are destroyed).

Litter interval – in this system it must be taken into account that half of the females whose litters are fostered will be returned for mating at the end of gestation and the other half will be returned after gestation and weaning, in both cases the length of the oestrous cycle must also be taken into account.

$$\text{Litter interval} = \frac{(\text{gestation} + \text{oestrous cycle}) + (\text{gestation} + \text{oestrous cycle} + \text{weaning})}{2} \text{ days}$$

$$= \frac{(22 + 5) + (22 + 5 + 21)}{2} \text{ days}$$

$$= \frac{27 + 48}{2} \text{ days}$$

$$= \frac{75}{2} \text{ days}$$

$$= 37.5 \text{ days (assume 38 days)}$$

$$\text{Number of litters per ♀ per year} = \frac{\text{number of days in year}}{\text{litter interval in days}}$$

$$= \frac{365}{38}$$

$$= 9.6 \text{ (assume 9)}$$

$$\text{Number of young per ♀ per year} = \text{number of litters per year} \times \text{number in litter}$$

$$= 9 \times 8$$

$$= 72$$

$$\text{Number of young per } ♀ \text{ per week} = \frac{\text{number of young per year}}{\text{number of weeks in year}}$$

$$= \frac{72}{50}$$

$$= 1.4 \text{ (assume 1)}$$

To produce 400 males each week a total of 800 young would need to be born each week (sex ration 1 ♀:1♂)

$$\text{Number of breeding females required} = \frac{\text{number of young required per week}}{\text{number of young per } ♀ \text{ per week}}$$

$$= \frac{800}{1}$$

800 breeding females will be required.

The economic breeding life of a rat is 6 litters, the litter interval is 38 days.

$$\text{Time to produce 6 litters} = 6 \times 38 \text{ days}$$

$$= 228 \text{ days or 32 weeks}$$

Therefore every week 1/32 of the breeding females should be replaced

$$= \frac{1}{32} \text{ of } 800$$

$$= \frac{800}{32}$$

$$= 25$$

25 females should be replaced each week. The young females can be reserved each week and fostered – as calculations have always been made to the advantage of the programme (p. 181) it would not be necessary to breed for this extra number. Males should be replaced at about 8 months of age; about 8 would be required each month and can be taken from the stock produced.

To produce 30 NZW rabbits per month

Relevant data	Range	Figure for calculation
Gestation	30–32 days	32 days
Litter size	6–8	6
Age of weaning	6 weeks	6 weeks 42 days

Breeding system – females mated 21 days after parturition
Litter interval = (length of gestation + time between parturition and mating) days
= (32+ 21) days
= 53 days

$$\text{Number of litters per ♀ per year} = \frac{\text{number of days in year}}{\text{litter interval}}$$

$$= \frac{365}{53}$$

$$= 6.88 \text{ (assume 7)}$$

$$\text{Number of young per ♀ per year} = \text{number of litters} \times \text{number in litter}$$

$$= 7 \times 6$$

$$= 42 \text{ young}$$

$$\text{Number of young per ♀ per month} = \frac{42}{12}$$

$$= 3.5 \text{ (assume 3)}$$

$$\text{Number of breeding females required} = \frac{\text{number required per month}}{\text{number of young per ♀ per month}}$$

$$= \frac{30}{3}$$

$$= 10$$

10 females are required.

The economic breeding life of the rabbit is 2 years so 1 female should be replaced every other month. 2 males would be required to mate this number of females and should be replaced one per year or when fertility falls, as shown by small litters or infertile matings.

Examination Technique 23

Theory

Presuming the relevant course has been attended and the syllabus covered, you can improve your performance in the examination by preparing yourself for the day, reading the instructions and the questions on the examination paper carefully, choosing your questions carefully, and planning your time in the examination.

Preparation for the day covers the time spent revising until the time of arrival at the examination centre. First make sure you know your subject, but do not spend the 2 or 3 days immediately prior to the examination revising from morning till night as you cannot possibly absorb all you can read or write in such a concentrated effort; it is better to work for shorter periods over a longer time. Do not prepare too elaborately for a session; all you require is your books, lecture notes, pencils and a lot of paper. Pencils and paper are most important – as you read make notes, draw diagrams when appropriate, and become familiar with the words, shapes and concepts on which you will be examined. If a rabbit heart is an essential part of your subject, learn to draw it quickly and accurately – the last place you want to learn to draw a rabbit heart is in the examination room. Remember a picture is as good as a thousand words: label diagrams fully, note how many labels there should be and check them on your drawing. Draw large diagrams, so that with their labels they span the page. You can have all the paper you need in an examination. Now and then you will find names, lists or properties for which mere familiarity and understanding are not enough – these must be learnt. They can either be written on a card and carried around and learnt during any spare time or a mnemonic can be made up; for example SEEMILL Sucus Entericus contains Enterokinase, Maltase, Invertase, Lactase and Lipase. Before you start each revision period, practise answering a question in a given time either from a previous examination paper or one you have set yourself. The benefit of this is twofold; you will practise setting your thoughts down in an orderly manner and you will discover how much you can write in a given time. If the allotted time is 25 minutes and you have finished your question in 10 minutes you have not written enough facts. On the other hand if you are only half way through the question after 25 minutes you are either putting in too much detail or waffling – either way you will not achieve a high mark. Check the answers later or ask your tutor or senior technician to do so. During revision you will have a lot of work to do, but be careful. If you are completely exhausted on the day after the examination you will have been nearly exhausted on the day – the very day you need to be 100% alert. It is probably useless to revise the evening before the examination – time would be better spent relaxing or going to bed early. There are some things that should be avoided the day prior to the examination. Don't sunbathe, as a sunburnt back can be a terrible distraction. So too can be an upset stomach, so beware of unusual foods. Needless to say an alcohol-sodden brain is a considerable disadvantage on such an occasion.

You will have details of where and at what time to attend the examination centre so you can arrange what bus or train to catch or if you are driving ensure you have sufficient petrol and know where to park; you do not want the worry of looking up timetables, or searching for a garage or parking space on the morning of the examination. It is your responsibility to present yourself on time; if you are late you may not be allowed to sit the examination.

At the start of the examination read the paper carefully. Read the instructions. How many questions are to be answered? Are any compulsory? Are questions to be answered from more than one section? How much time do you have for each question? Read all the questions, decide which you will attempt and make a note of the numbers. Start with the questions you feel most confident of answering. A calculation is clear and full marks can be awarded. Before starting a calculation get a 'common sense answer' by rounding off the figures given, for example 'How many mice will be produced by 2 females each having 4.8 litters, both with an average litter size of 9.7 young?' 4.8 is nearly 5 and 9.7 nearly 10 so $2 \times 5 \times 10 = 100$. This tells you the correct answer will be nearer 100 than 10 or 1,000 (errors of tenfold are very easy to make). Beware of questions that start 'Write an essay on . . .' or 'Give an account of . . .' as you will not receive full marks and could well run over the time allotted: you need to have your thoughts on the subject very well ordered for that sort of question. Questions that ask for brief notes are easier to answer in the time allowed and providing you write clearly and concisely, long sentences are not called for. Questions that are divided into parts will also gather marks even if you do not know the answers in depth.

Planning your time is most important. You should divide your time into two parts, one to give the main body of your response, the other for tidying it up. The following example illustrates how you should plan your time before you start the examination. Suppose you have six questions to answer in three hours: that gives you half-an-hour for each question. Allow less than this for planning and actually writing each answer – say twenty-five minutes. Spend the first five minutes reading the entire examination paper. So, starting the examination at 10.00 you should be starting the first question at 10.05 and a new question at the following times: 10.30, 10.55, 11.20, 11.45 and 12.10. It is most important that you adhere to this timetable; when your twenty-five minutes is up stop writing on that question whether you have finished or not and especially if you are less than half way through (the other half will take a disproportionately long time). If you have practised doing old examination questions under examination conditions during your revision you will find the real examination much easier as far as planning your time and setting down your thoughts is concerned. If you finish a question early write down a new timetable, still allowing twenty-five minutes for each remaining question. When you have done your six questions you will have at least 25 minutes left. You should use this to go through your paper picking up marks where you can. You can use it to finish questions, tidy up diagrams or check calculations. The required number of questions to be answered should always be attempted as each carries a set proportion of marks and a question not attempted at all is a total loss of the marks allocated for it. To attempt more than the required number is a waste of time as extra answers will be discarded.

Finally, remember marks can only be awarded for what the examiner can see, so display your thinking on paper; if the examiner can see how you got the wrong answer you could still get a pass mark for what your argument or calculations contained, so do not do mental arithmetic or make calculations or notes on scrap paper.

Practical

At the present time the practical examination of the Institute of Animal Technicians is divided into a section which may include dissection or microscopy techniques, and an oral section where several examiners ask questions on various aspects of laboratory animal husbandry, including handling and sexing animals.

Much of what has been said for the theory examination is relevant to the practical section; you need to know your subject and have practised the techniques you may be asked to demonstrate. Planning your time may be very important if, for example, the technique involved requires a period of time for staining or fixing material. You cannot afford to sit and watch for 10–15 minutes so make a note of the time of the start of the procedure and the time it is due to end so that in the meantime you can continue working on some other part of the examination.

You may be asked to dissect an animal and draw and label a specified system. If you are asked for the urinogenital system you are wasting your time if you include in your display the thoracic cavity. You will not gain marks if you draw and label what is not shown in your dissection; 2 asymmetrical kidneys and ureters look fine on a well practised diagram, but the animal in the examination did not evolve from a textbook – so draw what you actually see, carefully. Take care of your finished dissection by covering it with damp tissues or cotton wool so that it does not dry out before being marked by the examiner.

Keep your part of the bench or table clean and tidy, mop up any spillage as a blood- or chemical-stained answer paper can be difficult to read. All slides or solutions that are required by the examiner should be labelled and put to one side; slides left under the microscope may be broken or discarded by the next user and unlabelled solutions can easily be tipped away in error. At the end of the examination, lay your work out in order on the bench for marking.

Many students dread the oral section of the examination but this is unjustified as the examiner is only there to discover what you know and, once the conversation starts, the allotted time passes very quickly. When the examiner asks a question he should be treated as though he knows nothing about the subject, for example if he asks, 'How and where would you take blood from a rabbit', and you reply, 'With a syringe from its ear', you will receive fewer marks than if you briefly describe the preparation and holding of the rabbit, and in more detail the procedure for withdrawing the blood. You may be asked about a species you have no experience with – say so and then use your common sense as the basic requirements for all animals are similar. If at your place of work two 4-day-old calves arrived unexpectedly you would not leave them in the delivery van – a makeshift pen could be made with straw bales, so if asked about calves don't look at the examiner and say, 'Don't know, never worked with them,' because you will receive no marks for such an answer.

In the animal handling section you will be examined on your skill and knowledge of handling, sexing and determining age and weight of various species. When you handle the animals remember they are probably more nervous than you are. Handle them the way you do at work, confidently and firmly, and if you do have the misfortune to drop an animal, this is not instant failure. Act as you would at work, without fuss – pick up the animal, check for injury, then continue with the examination.

If you work with only one or two species, visit other animal units so that you can learn how others are housed and handled. A few days spent in other units will be more rewarding than just reading. There is no excuse for calling an adult hamster a baby guinea-pig.

Technician Education Council

In time it is intended that all technician education will be achieved by courses organised through the Technician Education Council (TEC, in Scotland SCOTEC).

TEC has established four basic qualifications at two levels: Certificate and Diploma and Higher Certificate and Higher Diploma. The Certificate equates roughly with the old ONC and the Higher Certificate with the HNC. At each level the Diploma embodies and broadens the Certificate.

The TEC courses are intended to train students to make them eminently employable within their field and also to develop them as people.

The main features of all schemes submitted to the TEC are that they must be shown to fit the needs of employers, often with a local emphasis and they are expected to educate the student as well as training him/her for a specific job. It is preferred that a certificate programme shall be arranged in units to give a total of 900 hours of study; many of the standard units would be common to other courses.

The basis of a TEC Life Sciences course is tripartite: (a) a basic scheme related to the subject, (b) the more obviously vocationally-related application, (c) the technology plus about 15% general and communication studies.

A typical 3-year programme for Animal Technicians would include units on: Safety and Laboratory Practice, Mathematics, Physical Sciences, Mammalian Physiology, Cell Biology, Biology, General and Communication Studies and of course Animal Technology.

Minimum entry qualification for such a course is three subjects at CSE Grade 3 or higher including a science subject and mathematics. Credits and exemptions from some units of study could be given to students who have passed related GCE examinations.

Appendix 1

Tables of practical information

1 Environment, caging and housing

	Room temp (°C)	Relative humidity %	Light duration (hours)	Cage size length×width×height (cm)	No. of animals per cage	Cage area per animal
Mouse	22	50–55	12–16	33×15×13	6	82 cm²
Rat	22	50–60	12–16	55×28×25	6	250 cm²
Guinea-pig	15–18	45–55	14	50×30×25	2	700 cm²
Syrian hamster	22	45–50	14–16	45×28×13	4	315 cm²
Chinese hamster	20	50	16	55×28×25	8	192 cm²
Gerbil	15–21	50	12–14	45×28×13	2	315 cm²
Rabbit	10–18	40–45	12–14	48×58×36	1	2784 cm²
Ferret	15	50	14–15	90×45×45	2	4050 cm²
Cat	15–22	40–45	14	90×63×50	1	2800 cm²
Dog	5–15	50	–	1800×1000	1	1.8 m²
				+ 2600×1000 outdoor run		+ 2.6 m²
Monkey (Macaque)	21–24	50–70	–	90×80×120	1	7200 cm²
Chicken	13	50	18	<2wks 50×25×25	20	65 cm²
				<6wks 50×30×30	5	320 cm²
Sheep	<15	–	–	1500×1200×1200	1	1.8 m²
Goat	<15	–	–	1800×1200×1300	1	2.2 m²
Pig	13–18	>80	–	3000×2400×1200	1 ♀ + young	7.2 m²
Cow	<15	–	–	2100×1000 cubicle	1	2.1 m²
Horse	<15	–	–	3600×3600×3000 loosebox	1	13.0 m²
				2200×1800×2000 stall		4.0 m²

2. Properties of Chemical Disinfectants

Disinfectant Group		Example or Brand Name	Cost *	ANTI-BACTERIAL ACTIVITY				INACTIVATION BY		
				Gram +	Gram –	Acid Fast	Spores	Hard Water	Organic Material	Synthetic Material
Alcohols		–	Moderate	Good	Good	Good	None	Slight	Slight	Slight
Aldehydes		Formaldehyde	Moderate	Good	Good	Good	Fair	Slight	Slight	Slight
		Glutaraldehyde – 'Cidex'		Good	Good	Fair	Good	–	Slight	Slight
Ampholytes		'Tego', 'Task'	Expensive	Good	Fair	None	None	Serious	Serious	Serious
Diguanides		Chlorhexidine – 'Hibitane'	Expensive	Good	Fair	Slight	None	Serious	Serious	Serious
Halogens	Hypochlorites	'Chloros', 'Domestos'	Cheap	Good	Good	Good	Good	Slight	Serious	Slight
	Iodine compounds	'Betadine', 'Pevidine' 'Wescodyne'		Good	Good	Good	Slight / Fair	Slight	Serious	Slight
	Chloroxylenol	'Dettol'	Expensive	Good	Fair	None	None	Serious	Serious	Moderate
	clear soluble	'Clearsol', 'Hycolin', 'Stericol', 'Sudol'	Moderate	Good	Good	Fair	None / Slight	Slight	Slight	Moderate
Phenolics	White Fluids	'Izal', 'White Cyllin'	Moderate	Good	Good	Fair	None / Slight	Slight	Slight	Moderate
	Black	'Jeyes Fluid'	Moderate	Good	Good	Fair	Slight	Slight	Slight	Moderate
Pine Fluids		'Jeypine', 'Pineodol'	Cheap	Slight	Slight	None	None	Serious	Serious	–
Quaternary Ammonium Compounds (QAC)	Cetrimide	'Cetavlon'	Expensive	Good	Fair	None	None	Serious	Serious	Serious
	Benzalkonium chloride	'Roccal'	Expensive	Good	Fair	None	None	Serious	Serious	Serious
QAC + Diguanides	Cetrimide chlorhexidine	'Savlon'	Expensive	Good	Fair	None	None	Serious	Serious	Serious
	Benzalkonium chloride + picloxydine	'Resiguard'	Expensive	Good	Fair	None	None	Serious	Serious	Serious

* Relative costs vary according to many factors, e.g. price of raw materials, bulk purchase discounts, etc.
Much of the information in this table has been taken from 'Hospital Hygiene', Pub. Arnold, London, by kind permission of the author, Isobel M. Maurer.

3. Laws relating to Animals

Cruelty to Animals Act 1976
Controls animal experiments

Protection of Animals Act 1911
Cruelty to animals in the sense of ill treatment

Protection of Animals Act 1934
Cruelty in the course of public entertainment

Protection of Animals (Anaesthetics) Act 1954
Defines the mandatory use of anaesthetics

Protection of Animals (Anaesthetics) Amendment Act 1964
Exceptions and additions to permitted procedures

Dogs Act 1871
Control and destruction of dangerous dogs

Dogs Act 1906
Strays and liability of owners. Prohibition of use of strays for research

Dogs (Amendment) Act 1928
Additions to and modifications of 1906 Act

Control of Dogs Order 1930
Wearing of collar with name and address disc and control in public places

Diseases of Animals Act 1950
Provision to make Orders to prevent the spread of disease. Lists notifiable diseases which must be reported to the police at the earliest opportunity. A very wide ranging Act

Dogs (Protection of Livestock) Act 1953
Sheep worrying and injury to livestock

Veterinary Surgeons Act 1948
Training and registration of veterinary surgeons. Defines procedures which may be performed by persons other than qualified veterinary surgeons

Veterinary Surgeons Act 1966
Repeals most of 1948 Act and modifies remainder particularly in respect of treatment permissible by lay persons

Laws relating to Farm Animals

Improvement of Livestock (Licensing of Bulls) Act 1931

Improvement of Livestock (Licensing of Bulls) Regulations 1955

Agriculture (Miscellaneous Provisions) Act 1944
Licensing of pedigree bulls

Licensing of Boars (England and Wales) Order 1949

Improvement of Livestock (Licensing of Boars) (England and Wales) Regulations 1956

Horse Breeding Act 1958
Licensing of stallions

Horse Breeding Rules 1948

Movement of Animals (Records) Order 1960
Defines records to be kept in 'movements book'

Live Poultry (Movement Records) Order 1958
Defines records to be kept by dealers, slaughtermen and auctioneers

Transit of Animals Order 1927
Sets out standards of accommodation and husbandry for transporting farm animal species by sea and rail in Great Britain

Animals (Sea Transport) Order 1930
As previous Order but for overseas transit

Transit of Animals (Amendment) Order 1931
Sets out standards of accommodation and husbandry for transporting animals by road vehicles

Transit of Calves Order 1963
Carriage of calves by road vehicles

Conveyance of Live Poultry Order 1919
Carriage by road, rail and sea

Transit of Horses Order 1951
Carriage of horses by road and rail

Horses (Sea Transport) Order 1952
Carriage of horses by sea

The Agriculture (Miscellaneous Provisions) Act 1968
The Brambell Report on the welfare of farm animals was published in 1965 and this Act three years later is of major importance in providing for the elimination of unnecessary pain or distress to farm livestock. 'Codes of Recommendations for the Welfare of Livestock' have been prepared in accordance with this Act

Poultry (Exposure for Sale) Order 1937

Markets (Protection of Animals) Order 1964

Docking and Nicking of Horses Act 1949

The Slaughter of Animals Act 1958

The Slaughter of Poultry Act 1967

Live Poultry (Restrictions) Order 1957 – 63
Poultry shows, sales and dealers. Movements to Scotland

Exportation of Animals

The Exported Animals Protection Order 1964
Conditions under which certain animals may be exported

The Ponies Act 1969
Export of ponies

Importation of Animals

Destructive Imported Animals Act 1932
Prohibited species and special import licences

Animals (Restriction of Importation) Act 1964
Licence required for importation of certain carnivores, marsupials and primates

Equine Animals (Importation) Order 1973
Licences and certificates of health, prohibitions at certain times of the year

The Rabies Order 1938
Sets out procedures to be followed if an animal is suspected of having rabies

The Importation of Dogs and Cats Order 1928–1970
Seeks to prevent the introduction of rabies carried by dogs and cats from abroad

The Exotic Animals (Importation) Order 1969
Revoked

The Rabies (Importation of Mammals) Order 1971
This important Order replaces or modifies the previous three Orders listed and its purpose is to prevent the introduction and spread of rabies into this country. Lists the species which can only be imported under licence and names the ports and airports through which such animals may enter. Sets out in detail the quarantine required and the supervision of the premises (which may be experimental animal units) and the method of transport to such premises. The Ministry of Agriculture, Fisheries and Food, Tolworth, Surrey issues licences for the importation of animals

Rabies Act 1974
Makes provision for Orders under the *Diseases of Animals Act 1950* to control an outbreak of rabies

Rabies (Importation of Dogs, Cats and other Mammals) Order 1974
Similar in principle to the *Rabies (Importation of Mammals) Order 1971*

4. Methods of Euthanasia

	Dislocate neck	Ether	Chloroform	CO_2	Barbiturates	Captive bolt humane killer
Mouse	●	●	●	●	–	–
Rat	●	●	●	●	–	–
Guinea-pig	●	–	●	●	–	–
Syrian hamster	●	●	●	●	–	–
Chinese hamster	●	●	●	●	–	–
Gerbil	●	●	●	●	–	–
Rabbit	●	–	–	●	●	–
Ferret	–	–	–	●	●	–
Cat	–	–	●	●	●	●
Dog	–	–	–	●	●	●
Monkey (Macaque)	–	–	–	–	●	–
Chicken	●	–	–	●	●	–
Sheep	–	–	–	–	●	●
Goat	–	–	–	–	●	●
Pig	–	–	–	–	●	●
Cow	–	–	–	–	●	●
Horse	–	–	–	–	●	●

● Suitable method

5. Formulae of commonly used diets

Diet 41B (rats, mice, monkeys)

	%
Ground wheat	47
Ground oats	40
White fishmeal	8
Dried skimmed milk	3
Dried brewers' yeast	1
Common salt	1
Vitamin supplement	

Diet PRM (rats, mice)

	Parts
Ground wheat	20
Middlings	20
Ground oats	20
Dried skimmed milk	12.5
Ground maize	10
Ground barley	5
Soya bean meal	10
White fishmeal	5
Dried brewers' yeast	0.5
Vitamin supplement	
Mineral supplement	

Diet RGP (rabbits, guinea-pigs)

	%
Ground barley	40
Middlings	15
Dried grassmeal	15
Ground oats	12.5
Linseed cake meal	10
White fishmeal	7.5
Vitamin supplement	
Mineral supplement	

Diet SG1 (rabbits, guinea-pigs)

	%
Middlings	58
Dried grassmeal	20
Ground oats	12
White fishmeal	10

When fed to guinea-pigs a supplement of vitamin C must be provided

In recent times diets have become standardised with respect to quality expressed as the calculated analysis; to achieve this a diet formula may be varied from batch to batch. The formulae here are given as a guide only to show typical ingredients.

6. Breeding Data

	Age at maturity		Weight at maturity		Age at sexual maturity		Age at first mating		Weight at first mating		Breeding system and mating ratio
	♂	♀	♂	♀	♂	♀	♂	♀	♂	♀	
Mouse	3 mths		40g	30g	5 wks		5 wks	6wks	34g	28g	pair, trio or 1:4
Rat	6 mths		500g	350g	40–50 days		10 wks		350g	250g	pair or harem 1:5
Guinea-pig	5 mths		700g	650g	60–80 days		12 wks		550g	500g	pair or harem 1:5
Syrian Hamster	8wks	9 wks	140g	110g	6wks		6wks		100g	80g	pair or harem ♀ taken to ♂ at oestrus
Chinese Hamster	3 months		50g	45g			10 wks		35g	40g	4:4, 2:6 if litter mates
Gerbil	6 months		85g	75g			18 wks	12 wks	75g	65g	monogamous pair
Rabbit	6–9 mths		4–6kg *		4–5 mths		5–6 mths		3–5kg *		♀ to ♂ at oestrus
Ferret	9 mths		1–3kg†		9 mths		6 mths		800g	750g	♀ to ♂ at oestrus
Cat	9 mths		3.5kg	2.5kg	8–9		7–9		3.3kg	2.5kg	
Dog	*		*		9–12 mths		14 mths		10–30kg *		
Monkey (Macaque)			9kg	4.5kg	3yrs	4yrs	5yrs	4.5yrs	1kg		pair or colony
Chicken	8–9 mths		*				8–9 mths		*		1:10
Sheep	14–18 mths		*				9 mths		*		‡
Goat	14–18 mths		*		3 mths		9–12 mths		*		‡
Pig	18 mths‡		*		4 mths	7 mths	8 mths		*		‡
Cow	2 yrs		*				1–1.5 yrs		*		usually artificial insemination
Horse	3 yrs		*								female to stud male

* Variable according to breed
† Ferret's body weight varies according to the time of year as fat is laid down or used up
‡ Variable

It should be noted that many of the figures quoted, especially ages and weights are average and may vary according to breed, strain, environment, litter size etc.

6. Breeding Data continued

	Economic breeding life	Breeding season	Ovulation mechanism	Length of oestrous cycle	Duration of oestrus	Gestation
Mouse	6 litters	continuous	spontaneous	4–5 days	10–20 hours	20 days
Rat	4–6 litters	continuous	spontaneous	5 days	10–20 hours	21 days
Guinea-pig	5–6 litters	continuous	spontaneous	14–16 days	6–12 hours	65–72 days
Syrian Hamster	1 yr	Feb–Oct	spontaneous	4 days	12 hours	16 days
Chinese Hamster	6 litters	continuous	spontaneous	4–5 days		20 days
Gerbil	3–7 litters	continuous	spontaneous	4–6 days	6 hours	24–26 days
Rabbit	2 yrs	continuous	induced	–	extended in	30–32 days
Ferret	3–4 yrs	March–Aug	induced	–	absence of male	42 days
Cat	6–8 yrs	continuous	induced	–	3–6 days	65 days
Dog	*	spring autumn	spontaneous	–	4–12 days	63 days
Monkey (Macaque)	>15 yrs	continuous	spontaneous	28 days	–	26 wks
Chicken	2 yrs	continuous	spontaneous	–	—	Incub. 21 days
Sheep	8–10 yrs	Sept–March	spontaneous	16–17 days	30–36 hours	147 days
Goat	8–10 yrs	Sept–Feb	spontaneous	21 days	24–36 hours	150 days
Pig	4–5 yrs	continuous	spontaneous	19–22 days	24–36 hours	108 days
Cow	8 yrs	continuous	spontaneous	21 days	3–6 hours	9 mths
Horse	20yrs	Feb–Aug	spontaneous	20–25 days	3–15 days	11 mths

6. Breeding Data continued

	Pseudo Pregnancy	Litter size	Average birth weight	Average weaning age	Average weaning weight	First oestrus after parturition
Mouse	12 days after sterile mating	6–10	1g	19 days	12g	immediately post parturition
Rat	12 days after sterile mating	6–10	5g	21 days	35g	immediately post parturition
Guinea-pig	–	3–4	90g	14 days	200g	immediately post parturition
Syrian Hamster	9–10 days	5–7	2g	21 days	40g	immediately post parturition
Chinese Hamster	–	4–8	1.5g	21 days	6g	end of lactation
Gerbil	14–16 days	3–5	3g	21 days	15g	immediately post parturition
Rabbit	17–19 days	6–12	*	5 wks	800–1000g	3rd week of lactation
Ferret	42 days	6–10	6–12g	7 wks	300–400g	end of lactation
Cat	23 days	3–5	110g	4–5 wks	300–500g	4th week of lactation
Dog	60 days	4–8	*	6–8 wks	*	next season
Monkey (Macaque)	–	1	500g	3 mths	700–1000g	2–3 months
Chicken	–	–	20–40g	–	–	–
Sheep	rare - 140–147 days	1–3	2.5–3kg	12 wks	18–22kg	next season
Goat	rare - 150 days	1–3	2.5–3.5kg	6–12 wks	20–30kg	next season
Pig	–	8–15	*	3–6	–	end of lactation
Cow	–	1–2	*	5–12 wks	–	early in lactation
Horse	–	1	*	6 mths	–	early in lactation

7. Physiological norms

	Body temp. °C	Respiration rate per min	Pulse rate per min	Life span years	Chromo-some number	Mean plasma volume ml/kg body wt.	Haemoglobin g/100ml blood	RBC 10^6/cc	WBC 10^3/cc
Mouse	37–37.5	85–230	320–780	2	40	49.2	13.57–16.21	6.4–11.5	8.0–11.0
Rat	37.5–38.5	66–150	261–600	3	42	31.3	12.0–17.5	6.6–9.7	7.6–14.0
Guinea-pig	38.3–39.0	69–150	130–400	4–5	46	38.6	11.0–17.2	4.5–7.0	3.2–15.0
Syrian hamster	36.7–37.4	33–127	101–600	2	44	47–50	2.0–30.0	6.84–9.23	3.4–7.6
Chinese hamster	37–37.5	–	–	3	22	–	15.8±0.98	7.5±2.4	6.05±1.66
Gerbil	37.4	70–120	260–600	2–3	44	–	13.9–14.4	7.7–8.7	8.3–11.2
Rabbit	38.3–39.7	36–50	123–304	5	44	38.8	8.0–15.0	4.6–5.7	5.2–12.0
Ferret	37.8–40.0	33–36	–	5–6	40	–		8.5	8.9
Cat	38.0	20–30	100–120	12	38		7.0–15.5	6.5–9.5	9.0–24.0
Dog	38.2–38.7	15–30	90–100	13	78	53	11.0–18.0	4.5–8.0	8.0–18.0
Monkey (Macaque)	36.0–40.0	30–49	150–333		42	44.7	10.0–16.0	3.6–6.8	8.0–25.0
Chicken	41.6	30–35	280–400	14	77 or 78	–	7.3–12.9	2.0–3.2	9.1–56.0
Sheep	38.4–41.0	10–25	70–80	10–12	54	61.9	10.0–11.8	9.4–11.1	5.0–10.0
Goat	38.4–41.0	10–20	70–80	10–12	60	55.9	8.8–11.4	13.3–17.9	5.0–14.0
Pig	38.2–40.5	10–16	70–80	10	40	41.9	13.2–14.2	6.4	7.0–20.0
Cow	38.7–39.0	12–16	45–50	25	38	38.8	8.7–14.5	6.1–10.7	6.0–12.0
Horse	38–38.2	8–12	36–42	30	66	51.0	8.0–14.0	8.2–10.4	5.0–11.0

8. Anatomical norms

	Number of nipples/ teats	Nipples/ teats present in male	Toes front	Toes hind	Dentition I	C	PM	M
Mouse	10	No	4	5	1	0	0	3
					1	0	0	3
Rat	12	No	4	5	1	0	0	3
					1	0	0	3
Guinea-pig	2	Yes	4	3	1	0	0	4
					1	0	0	4
Syrian hamster	12–14	No	4	5				
Chinese hamster	8	No	4	5				
Gerbil			5	4				
Rabbit	8	Yes	5	4	2	0	3	3
					1	0	2	3 or
					1	0	3	3
Ferret	8	Yes	5	5	3	1	2	2
					3	1	3	2
Cat		Yes	5	5	3	1	3	1
					3	1	2	1
Dog	8–10	Yes	4	4	3	1	4	2
					3	1	4	3
Monkey (Macaque)	2	Yes	5	5	2	1	2	3
					2	1	2	3
Chicken	–	–	–	4	–			
Sheep	2	Yes	2	2	0	0	3	3
					3	1	3	3
Goat	2	Yes	2	2	0	0	3	3
					4	0	3	3
Pig	12–16	Yes	2	2	3	1	4	3
					3	1	4	3
Cow	4	Yes	2	2	0	0	3	3
					4	0	3	3
Horse	2	No	1	1	3	1	0	6 or 7
					3	1	0	6 or 7

I = Incisor C = Canine PM = Premolar M = Molar

9. Food and water intake

	Food	Water
Mouse	5 g	6.7 ml
Rat	10 g	20.8 ml
Guinea-pig	8 g per 100 g bdy wt	14.5 ml per 100 g bdy wt
Syrian Hamster	10–15 g	8–12 ml
Chinese Hamster	4–5 g	5 ml
Gerbil	5–7 g	4 ml per 100 g bdy wt
Rabbit	†140 g	†150–200
Ferret		30 ml (wet diet)
Cat	150–200 g wet	7 ml (wet diet) 70 ml (dry diet)
Dog	†450 g	†1000 ml
Monkey (Macaque)	50 g per kg bdy wt	75 ml per kg bdy wt
Chicken – day old	5 g	34 ml
– 10 wks	150 g	160 ml
Sheep	*†Hay: concentrate 3:1 up to 1.5–2 kg day	* 0.5–12 litres
Goat	1500 g concentrate 1500 g hay *	4–18 litres
Pig	*	up to 20 litres *
Cow	27 kg silage or 9 kg hay	dry 33–45 litres lactating + 5 × vol of milk
Horse	2% bdy wt as dry matter	30–35 litres

* Depends on age, weight and productive status
† Depends on breed

10. Bedding and nesting materials

	Bedding	Nesting	Droppings tray/ beneath cage
Mouse Rat	softwood shavings, sawdust or flakes	shredded paper, wood wool, paper tissue	softwood shavings, sawdust, flakes, peat, attapulgite clay
Guinea-pig	softwood shavings, sawdust or flakes	hay	softwood shavings, sawdust, flakes, peat, attapulgite clay
Hamster	softwood shavings, sawdust or flakes	shredded paper, wood wool	softwood shavings, sawdust, flakes, peat, attapulgite clay

	Bedding	Nesting	Droppings tray / beneath cage
Gerbil		shredded paper, paper tissue	softwood shavings, sawdust, flakes, peat, attapulgite clay
Rabbit	softwood shavings, sawdust or flakes	hay, softwood shavings	
Ferret		hay	
Cat	–	paper, wood wool	litter tray, sawdust, peat, attapulgite clay
Dog	–	paper, softwood shavings	
Monkey	–	–	sawdust
Chicken	softwood shavings, woodflakes, straw	straw, hay	sawdust
Sheep Goat Pig Cow Horse	softwood shavings, woodflakes, straw	straw (dusty or fine materials such as shavings should not be used as the female may be discouraged from cleaning the young after parturition, except pigs – sows do not clean young)	

11. Identification methods

Mouse	punched hole ear code, tail tattoo, dye spots
Rat	punched hole ear code, tail tattoo, dye spots
Guinea-pig	ear tattoo, dye spots
Hamster	punched hole ear code, dye spots
Gerbil	punched hole ear code, dye spots
Rabbit	ear tattoo, ear tag, leg ring, dye spots
Ferret	ear tattoo, dye spots
Cat	collar and disc, coat colour pattern
Dog	collar and disc, ear tattoo
Monkey	tattoo on chest, upper lip, forehead or inside thigh, waist chain and disc
Chicken	wing tag, leg ring
Sheep	ear tattoo, ear tag
Goat	ear tattoo, ear tag, collar and disc
Pig	punched hole ear code, ear tag, ear tattoo
Cow	punched hole ear code, ear tag, neck band, freeze or hot branding
Horse	coat colour, freeze or hot branding

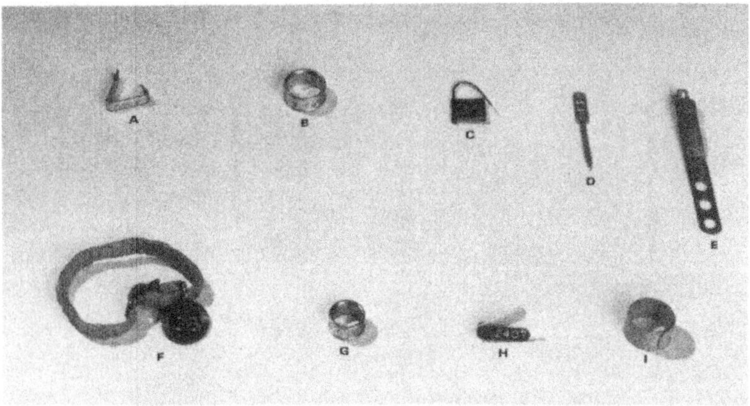

A11 *Identification tags (a) 'ketchum -tag' (b) rabbit leg ring (c) chicken wing tag (d) chicken wing tag (e) chicken leg ring (open) (f) cat collar (g) pigeon ring (h) rabbit 'roto -tag' (i) expanding chicken leg ring*

12. SI (International) Units, giving conversion factors

Length

metre (m) centimetre (cm) millimetre (mm)

1 m	= 100 cm	1 in	= 2.54 cm
1 cm	= 10 mm	1 ft	= 30.48 cm
1 cm	= 0.394 in	1 yd	= 0.914 m
1 m	= 3.281 ft	1 mile	= 1.61 km
	= 1.094 yd		
1 km	= 0.62 mile		

Area

1 m^2	= 10000 cm^2	1 in^2	= 6.452cm^2 = 645.2 mm^2
1 cm^2	= 0.155 in^2	1 ft^2	= 0.09 m^2
1 m^2	= 10.746 ft^2	1 yd^2	= 0.836 m^2

Volume

1 m^3	= 1,000,000 cm^3	1 in^3	= 16.39 cm^3
1 cm^3	= 0.061 in^3	1 ft^3	= 0.03 m^3
1 m^3	= 35.31 ft^3	1 yd^3	= 0.765 m^3
1 dm^3	= 1000 cm^3	1 pint	= 0.568 litres
	= 1 litre		(dm^3) = 568 cm^3
		1 gallon	= 4.546 litres (dm^3)
1 litre	= 1.76 pints	1 (US) gallon	= 3.785 litres (dm^3)
	= 0.22 gallons		= 0.83 (GB) gallon
	= 0.035 ft^3	1 ft^3	= 6.23 gallons
	= 35.19 fluid ounces	1 gallon	= 0.16 ft^3
		1 fluid ounce	= 28.4 cm^3

Mass

tonne (t) kilogramme (kg) gramme (g)

1 t	= 1000 kg	1 grain	= 0.0143 lb
1 kg	= 1000 g		= 0.0648 g
1 g	= 15.4 grains	1 oz	= 28.4 g
	= 0.035 oz	1 lb	= 453.6 g
1 kg	= 2.205 lb	1 ton	= 20 cwt = 80 quarters
1 t	= 0.984 ton		= 160 stones = 2240 lb
			= 1.017 t

Pressure

Pascal (Pa) Newton (N)

1 Pa = 1 N/m^2
1 atmosphere = 1.013×10^5 Pa
1 mm Hg (mercury) = 133.3 Pa = 0.133 kPa
1 lbf/in^2 (psi) = 6895 Pa = 6.895 kPa
1 lbf/ft^2 = 47.88 Pa

1 atmosphere	= 760 mm Hg	1 lbf/in^2	= 70.3 g/cm^2
	= 29.92 inHg		= 0.0608 atmospheres
	= 14.70 lbf/in^2		= 2.042 inHg
	= 33.89 ft water		= 51.8 mmHg
			= 27.71 in water
			= 2.31 ft water

Energy (heat, work)

Joule (J)

1 joule = 0.239 cal
1 cal = 4.19 J
1 k cal = 1Cal = 1000 cal
= 4.19 kJ

Temperature

degrees Celsius (°C)

1 degree Celsius = 1 degree Centigrade
0°C = 32°F
37°C = 98.6°F (human body temperature)

13. Conversion Tables for Metric and Imperial Measurement

Length			Area			Volume			Mass		
ins		**mm**	**sq in**		**cm^2**	**cu in**		**cm^3**	**oz**		**g**
0.039	**1**	25.4	0.155	**1**	6.452	0.061	**1**	16.387	0.035	**1**	28.350
0.079	**2**	50.8	0.310	**2**	12.903	0.122	**2**	32.774	0.071	**2**	56.699
0.118	**3**	76.2	0.465	**3**	19.355	0.183	**3**	49.161	0.106	**3**	85.048
0.157	**4**	101.6	0.620	**4**	25.806	0.244	**4**	65.548	0.141	**4**	113.398
0.197	**5**	127.0	0.775	**5**	32.258	0.305	**5**	81.936	0.176	**5**	141.748
0.236	**6**	152.4	0.930	**6**	38.710	0.366	**6**	98.323	0.212	**6**	170.097
0.276	**7**	177.8	1.085	**7**	45.161	0.427	**7**	114.710	0.247	**7**	198.446
0.315	**8**	203.2	1.240	**8**	51.612	0.488	**8**	131.097	0.282	**8**	226.796
0.354	**9**	228.6	1.395	**9**	58.064	0.549	**9**	147.484	0.317	**9**	255.146
ft		**m**	**sq ft**		**m^2**	**cu ft**		**m^3**	**lb**		**kg**
3.281	**1**	0.305	10.764	**1**	0.093	35.315	**1**	0.028	2.205	**1**	0.454
6.562	**2**	0.610	21.528	**2**	0.186	70.629	**2**	0.057	4.409	**2**	0.907
9.843	**3**	0.914	32.292	**3**	0.279	105.944	**3**	0.085	6.614	**3**	1.361
13.123	**4**	1.219	43.056	**4**	0.372	141.259	**4**	0.113	8.818	**4**	1.814
16.404	**5**	1.524	53.820	**5**	0.465	176.573	**5**	0.142	11.023	**5**	2.268
19.685	**6**	1.829	64.583	**6**	0.557	211.888	**6**	0.170	13.228	**6**	2.722
22.966	**7**	2.134	75.347	**7**	0.650	247.203	**7**	0.198	15.432	**7**	3.175
26.247	**8**	2.438	86.111	**8**	0.743	282.517	**8**	0.227	17.637	**8**	3.629
29.528	**9**	2.743	96.875	**9**	0.836	317.832	**9**	0.255	19.842	**9**	4.082
yds		**m**	**sq yd**		**m^2**	**pints**		**litres**	**tons**		**tonnes**
1.094	**1**	0.914	1.196	**1**	0.836	1.760	**1**	0.568	0.984	**1**	1.016
2.187	**2**	1.829	2.392	**2**	1.672	3.520	**2**	1.137	1.968	**2**	2.032
3.281	**3**	2.743	3.588	**3**	2.508	5.279	**3**	1.705	2.953	**3**	3.048
4.374	**4**	3.658	4.784	**4**	3.345	7.039	**4**	2.273	3.937	**4**	4.064
5.468	**5**	4.572	5.980	**5**	4.181	8.799	**5**	2.841	4.921	**5**	5.080
6.562	**6**	5.486	7.176	**6**	5.017	10.559	**6**	3.410	5.905	**6**	6.096
7.655	**7**	6.401	8.372	**7**	5.853	12.318	**7**	3.978	6.889	**7**	7.112
8.749	**8**	7.315	9.568	**8**	6.689	14.078	**8**	4.546	7.874	**8**	8.128
9.843	**9**	8.230	10.764	**9**	7.525	15.838	**9**	5.114	8.858	**9**	9.144

acres		hectares	gallons		litres
2.471	**1**	0.405	0.220	**1**	4.546
4.942	**2**	0.809	0.440	**2**	9.092
7.413	**3**	1.214	0.660	**3**	13.638
9.884	**4**	1.619	0.880	**4**	18.184
12.355	**5**	2.023	1.100	**5**	22.730
14.826	**6**	2.428	1.320	**6**	27.277
17.297	**7**	2.833	1.540	**7**	31.823
19.769	**8**	3.237	1.760	**8**	36.369
22.240	**9**	3.642	1.980	**9**	40.915

The bold figures in the central columns can be read as either metric or British measure. Thus 1 in = 25.4 mm or 1 mm = 0.039 in.

14. Temperature Conversion Table for Fahrenheit and Celsius (Centigrade)

°F		°C	°F		°C	°F		°C	°F		°C
−4.0	**−20**	−28.9	71.6	**22**	−5.6	147.2	**64**	17.8	222.8	**106**	41.1
−2.0	**−19**	−28.3	73.4	**23**	−5.0	149.0	**65**	18.3	224.6	**107**	41.7
−0.4	**−18**	−27.8	75.2	**24**	−4.4	150.8	**66**	18.9	226.4	**108**	42.2
1.4	**−17**	−27.2	77.0	**25**	−3.9	152.6	**67**	19.4	228.2	**109**	42.8
3.2	**−16**	−26.7	78.8	**26**	−3.3	154.4	**68**	20.0	230.0	**110**	43.3
5.0	**−15**	−26.1	80.6	**27**	−2.8	156.2	**69**	20.6	231.8	**111**	43.9
6.8	**−14**	−25.6	82.4	**28**	−2.2	158.0	**70**	21.1	233.6	**112**	44.4
8.6	**−13**	−25.0	84.2	**29**	−1.7	159.8	**71**	21.7	235.4	**113**	45.0
10.4	**−12**	−24.4	86.0	**30**	−1.1	161.6	**72**	22.2	237.2	**114**	45.6
12.2	**−11**	−23.9	87.8	**31**	−0.6	163.4	**73**	22.8	239.0	**115**	46.1
14.0	**−10**	−23.3	89.6	**32**	0	165.2	**74**	23.3	240.8	**116**	46.7
15.8	**−9**	−22.8	91.4	**33**	0.6	167.0	**75**	23.9	242.6	**117**	47.2
17.6	**−8**	−22.2	93.2	**34**	1.1	168.8	**75**	24.4	244.4	**118**	47.8
19.4	**−7**	−21.7	95.0	**35**	1.7	170.6	**77**	25.0	246.2	**119**	48.3
21.2	**−6**	−21.1	96.8	**36**	2.2	172.4	**78**	25.6	248.0	**120**	48.9
23.0	**−5**	−20.6	98.6	**37**	2.8	174.2	**79**	26.1	249.8	**121**	49.4
24.8	**−4**	−20.0	100.4	**38**	3.3	176.0	**80**	26.7	251.6	**122**	50.0
26.6	**−3**	−19.4	102.2	**39**	3.9	177.8	**81**	27.2	253.4	**123**	50.6
28.4	**−2**	−18.9	104.0	**40**	4.4	179.6	**82**	27.8	255.2	**124**	51.1
30.2	**−1**	−18.3	105.8	**41**	5.0	181.4	**83**	28.3	257.0	**125**	51.7
32.0	**0**	−17.8	107.6	**42**	5.6	183.2	**84**	28.9	258.8	**126**	52.2
33.8	**1**	−17.2	109.4	**43**	6.1	185.0	**85**	29.4	260.6	**127**	52.8
35.6	**2**	−16.7	111.2	**44**	6.7	186.8	**86**	30.0	262.4	**128**	53.3
37.4	**3**	−16.1	113.0	**45**	7.2	188.6	**87**	30.6	264.2	**129**	59.9
39.2	**4**	−15.6	114.8	**46**	7.8	190.4	**88**	31.1	266.0	**130**	54.4
41.0	**5**	−15.0	116.6	**47**	8.3	192.2	**89**	31.7	267.8	**131**	55.0
42.8	**6**	−14.4	118.4	**48**	8.9	194.0	**90**	32.2	269.6	**132**	55.6
44.6	**7**	−13.9	120.2	**49**	9.4	195.8	**91**	32.8	271.4	**133**	56.1
46.4	**8**	−13.3	122.0	**50**	10.0	197.6	**92**	33.3	273.2	**134**	56.7
48.2	**9**	−12.8	123.8	**51**	10.6	199.4	**93**	33.9	275.0	**135**	57.2
50.0	**10**	−12.2	125.6	**52**	11.1	201.2	**94**	34.4	276.8	**136**	57.8
51.8	**11**	−11.7	127.4	**53**	11.7	203.0	**95**	35.0	278.6	**137**	58.3
53.6	**12**	−11.1	129.2	**54**	12.2	204.8	**96**	35.6	280.4	**138**	58.9
55.4	**13**	−10.6	131.0	**55**	12.8	206.6	**97**	36.1	282.2	**139**	59.4
57.2	**14**	−10.0	132.8	**56**	13.3	208.4	**98**	36.7	284.0	**140**	60.0
59.0	**15**	−9.4	134.6	**57**	13.9	210.2	**99**	37.2	285.8	**141**	60.6
60.8	**16**	−8.9	136.4	**58**	14.4	212.0	**100**	37.8	287.6	**142**	61.1
62.6	**17**	−8.3	138.2	**59**	15.0	213.8	**101**	38.3	289.4	**143**	61.7
64.4	**18**	−7.8	140.0	**60**	15.6	215.6	**102**	38.9	291.2	**144**	62.2
66.2	**19**	−7.2	141.8	**61**	16.1	217.4	**103**	39.4	293.0	**145**	62.8
68.0	**20**	−6.7	143.6	**62**	16.7	219.2	**104**	40.0	294.8	**146**	63.3
69.8	**21**	−6.1	145.4	**63**	17.2	221.0	**105**	40.6	296.6	**147**	63.9

The figures in bold type indicate the temperature (°F or °C) to be converted.
For example to find the Celsius equivalent of 68°F look up 68 in bold figures and read off the equivalent in the Celsius column namely 20°C

The formulae for converting °F to °C is: $\frac{5\,(F-32)}{9}$ and for °C to °F: $\frac{9\,(C)}{5} + 32$

where
F = Temperature in degrees Fahrenheit
C = Temperature in degrees Celsius (Centigrade)

15. Mathematical signs and abbreviations

+	plus
−	minus
×	multiply
÷	divide
=	equals
≠	is not equal to
≑	is approximately equal to
>	greater than
<	less than
∴	therefore
∴	because or since
%	per cent
//	parallel to
⊥	perpendicular to
△	triangle
□	square
▭	rectangle
○	circle
∠	angle
∟	right angle
√	square root
°	degree
≃	is approximately equal to
⩾	is greater than or equal to
⩽	is less than or equal to
Σ	the sum of
~	the difference between
∝	varies as
#	number

Appendix 2
Glossary

abscess – a pathological lesion usually filled with pus
acute disease – a condition which appears rapidly and either leads rapidly to death or recovery
alopecia – loss of hair
anaesthetic – a drug administered to animals to prevent feeling pain, can be local or general – latter renders animal unconscious
anodising – producing an oxide layer on metal to form a protective coating (usually to aluminium)
anoestrus – a period of sexual inactivity in female animals
anorexia – inability or refusal to eat
antiseptic – a substance applied to living tissue to prevent sepsis or putrefaction
antisera – sera containing specific antibodies
aseptic – bacteria excluded
athymic – without a thymus gland (e.g. nude mouse)
barbiturates – group of drugs used for anaesthesia and euthanasia; all of those in clinical use are sodium salts of barbituric acid or its thio-derivative eg sodium pentobarbitone
bi-metallic strip – two strips of metal welded or riveted together to form a metal 'sandwich' which bends when heated due to the different expansion of each metal
brachycardia – abnormally slow heart beat and thus pulse rate
Browne's tubes – indicator tubes filled with heat sensitive liquid which are used as a temperature check in autoclaves
calculi – stones containing various salts found in various parts of the body e.g. kidneys and bladder. Can cause fatal blockages
carcinogen – a substance that will cause the growth of neoplasms (cancer)
caudal – tail region; the caudal vertebrae are those in the tail
chronic – a condition that is long lasting and persistent. In the case of disease or infection it may resolve but not always
cilia – microscopic whip-like processes covering certain cells as on certain mucous membranes and some micro-organisms
clinical sign – the sign of disease or infection that can be detected by observation or simple diagnostic procedures
cloaca – common opening for reproduction system and excretion from alimentary canal in birds

closed colony – an animal breeding colony into which no new animals are introduced, all future stock coming from within the colony
congenic – having similar genotypes
congenital – traits, usually undesirable, that are passed on from parent to offspring e.g. deformities
conjunctivitis – inflammation of the conjunctiva, the membrane covering the front of the eye
coprophagy – the eating of its faeces by an animal, a normal practice in rabbits
copulation – the act of mating, coitus, sexual intercourse
curare – a drug causing paralysis by complete relaxation of all muscle. It does not prevent an animal feeling pain. For use in animal experiments, Home Office permission must be granted
damper – a device for slowing down movement. Used on balances to slow down movement of scale pan, e.g. when weighing animals
DNA – deoxyribonucleic acid, a complex chemical constituent of chromosomes in the nucleus which stores genetic information
dorsal – along the back or above, opposite to ventral
dyspnoea – difficulty in breathing
dystokia or dystocia – difficulty during parturition
embolism – something that blocks a small blood vessel e.g. fat, bacteria or an air bubble
endemic – commonly found, as in the case of a disease that is always present in a population
enteritis – inflammation of the intestine
enzootic – an outbreak of a disease in particular areas. The disease becomes epizootic if it spreads outside the particular area
epidemic – a disease that becomes prevalent in a population at a specific time
epizootic – a term applied to a disease which affects animals over a wide area and which spreads rapidly
equidae – members of the horse family
euthanasia – humane killing
exothermic – a process in which heat is produced
fetus – unborn offspring, a developed embryo
fomite – an inanimate object on which pathogens can be carried
galvanised – metal that has been coated with a thin layer of zinc (zinc plated) to prevent corrosion
genitalia – external organs of reproduction
genotype – the genetic composition of an animal; may describe the genetic background of a single trait or a number of traits
gnotobiotic – literally means 'known life', usually used to describe germ free animals
haemorrhage – bleeding
harem – a group of females served by one male
heterozygous – a genetic term describing a character with different alleles at a given locus e.g. Aa is heterozygous, AA is not
homozygous – a genetic term describing a character having identical alleles at a given locus. e.g. AA is homozygous but Aa is not
hybrid – the progeny of a mating between animals of different genetic background

hysterectomy – removal of the uterus
hysterotomy – the opening of the uterus (caesarean section)
inorganic substance – produced from minerals and not containing carbon
isolator – a metal or plastic container in which animals are kept totally isolated from the outside environment. Access is via sterilisation ports. Isolators are used for the production of germ free animals
lesion – an injury to an organ or living tissue
lordosis – abnormal curvature of the spine; may be pathological, may be shown by females as a sign of receptiveness to a male
macroscopic – large, visible to the naked eye
menstrual cycle – the ovulation cycle of primates, which is accompanied by bleeding when the endometrium of the uterus breaks down if fertilization does not occur
mesentery – the layer of tissue that supports the intestine in the peritoneum
metastasis – a secondary tumour that occurs away from the primary malignant growth
microscopic – small, usually can only be seen through a microscope
moribund – dying or near to death
mycoplasma – an infective agent that is neither virus nor bacteria. Similar in size to a large virus.
necropsy – examination of internal structures after death
neoplasm – a tumour
New World monkeys – those found in Central and South America
oestrous cycle – the ovulation cycle of mammals
oestrus – the period in the sexual (oestrous) cycle when a female is receptive to a male
Old World monkeys – those from Asia, the Far East and the African continent
oocysts – reproductive cysts usually associated with internal parasites e.g. coccidia
organic substance – produced from living organisms and containing carbon
palatable – pleasant to taste
paralysis – inability of muscles to function
parturition – giving birth
pasteurization – application of heat to kill specific micro-organisms or to reduce their efficacy
pathogen – a disease-causing micro-organism
pathology – the science of studying the changes in a body brought about by disease
peritoneum – the membrane lining the abdominal cavity
plasma – the complete fluid portion of blood collected by centrifugation
poly – many
polycarbonate – a transparent plastic, tough, heat distortion 135°C
polyethylene – a soft translucent plastic usually called polythene, heat distortion 82°C
polypropylene – a translucent plastic commonly used for animal cages, heat distortion 105°C–110°C
polystyrene – a brittle plastic often used for disposable containers, heat distortion 65°C–75°C

prehensile – capable of gripping e.g. the tail of New World monkeys
prill – a small bead or granule
protozoa – single celled organisms e.g. coccidia
pseudopregnant – false pregnancy usually after a sterile mating. Pseudopregnancy is usually shorter than a true pregnancy although outward signs may be similar
quarantine – isolation to prevent infection
RNA – ribonucleic acid, a complex chemical which transmits genetic information from the nucleus to the cytoplasm where it controls the production of the right chemical at the right time
sedation – the administration of a drug to reduce activity e.g. when handling monkeys
septic – putrefying (as in an infected lesion)
serum – the non-cellular fluid portion of clotted blood
sternum – the breast bone
stools – faeces
sub-clinical disease – a disease where no signs are apparent
thermocouple – an electrical device sensitive to temperature. Sometimes connected to a dial to give a recording of temperature
torsion bar – a metal bar that resists a force by twisting. Often used in balances to resist movement of loaded pan to give a graduated reading via a pointer
vector – a carrier, used to describe a carrier of infectious micro-organisms
ventral – underneath, opposite of dorsal
zoonotic, zoonosis – disease or infection that is transmissible between animals and man, e.g. rabies

Appendix 3
Suggested further reading

Environment

Clough, G & Gamble M R (1976) *A guide to the design and planning of animal facilities*, LAC Manual Series No. 4. Medical Research Council Laboratory Animals Centre.

Hare, R & O'Donoghue, P N eds. (1968) *The design and function of laboratory animal houses*. Laboratory Animal Handbooks 1. Laboratory Animals, London.

LAC Collected Papers Vol 11 (1962) *The environment of laboratory animals*. Medical Research Council Laboratory Animals Centre.

McSheehy, T ed. (1976) *Control of the animal house environment*. Laboratory Animal Handbooks 7. Laboratory Animals, London.

Sainsbury, D W D (1967) *Animal health and housing*. Ballière, Tindall & Cox, London.

Hygiene

Kelsey, J C (1969) Disinfectants and methods of testing. *J. Medical Laboratory Technology* **26**, 79.

Kelsey, J C & Maurer, I M (1966) An in-use test for hospital disinfectants. *Monthly Bulletin of the Ministry of Health and the Public Health Laboratory Service* **25**, 180.

Maurer, I M (1976) *Hospital Hygiene*. Arnold, London.

Passmore, R & Robson, J S (1980) *A companion to medical studies*. Vol. 2, Ch. 18. (Microbiology, sterilization & disinfection) Blackwell scientific publications, Oxford.

Sykes, G (1965) *Disinfection and sterilization*. Spon, London.

Transport

British Standards Institution (1961–66) BS 3149: *Recommendations for carriage of live animals by air*. Parts 1–10.

King, J O L (1970) *Transport of animals*. UFAW, Potters Bar.

UFAW (1975) *Transport of farm animals*. UFAW, London.

Law

Cook, A J D (1976) *a Guide to laboratory law*. Butterworths, London.

Littlewood (1965) *Report of the departmental committee on experiments on animals*. HMSO, London.

North, P M (1972) *The modern law of animals.* Butterworths, London.
R D S (1979) *Notes on the law relating to experiments on animals in Great Britain.* 4th edn. Research Defence Society, London.
Sandys-Winsch, G. (1978) *Animal law.* Shaw, London.
Thomas, J L (1975) *Diseases of animals law.* 7th edn. Police Review, London.
UFAW (1975) *Animals and the law.* UFAW, London.
Wyatt, H V ed. (1980) *Handbook for the animal licence holder.* Institute of Biology, London.

Nutrition

Agricultural Research Council (1965) *The nutrient requirements of farm livestock* No. 2, Ruminants. Technical reviews and summaries. HMSO, London.
Agricultural Research Council (1966) *The nutrient requirements of farm livestock* No. 3, Pigs. Technical Review, ARC, London.
Evans, R E (1960) *Rations for livestock.* MAFF Bulletin No.48, HMSO, London.
LAC (1977) *Dietary standards for laboratory animals (report of the laboratory animals Centre Diets Advisory Committee)* Medical Research Council, Laboratory Animals Centre.
Nutrient requirements of rabbits (1977). 2nd edn. National Academy of Sciences, Washington.
Tavenor, W D ed. (1970) *Nutrition and disease in experimental animals.* Baillière, Tindall & Cassell, London.

Breeding & Reproduction

Arthur, G H (1964) *Wrights veterinary obstetrics*, 3rd edn. Baillière, Tindall & Cox, London.
Bowman, C J (1974) *An introduction to animal breeding.* Arnold, London.
Hafez, E S E (1970) *Reproduction and breeding techniques for laboratory animals.* Lea & Febiger, Philadelphia.
Lane-Petter, W ed. (1963) *Breeding and management of some common laboratory animals.* Academic Press, London.
Perkins, F T & O'Donoghue, P N eds. (1975) *Breeding simians for developmental biology.* Laboratory Animal Handbooks 6. Laboratory Animals. London.
Porter, G & Lane-Petter, W eds. (1962) *Notes for breeders of common laboratory animals.* Academic Press, London.
Pugh, R (1968) *The mouse, its reproduction and development.* Burgess, Minneapolis.

Specialised Techniques

Boddie, G F (1969) *Diagnostic methods in veterinary medicine.* 6th edn. Oliver & Boyd, Edinburgh.
Bush, B M (1975) *Veterinary laboratory manual.* Heinemann, London.
Coates, M E ed. (1968) *The germ free animal in research.* Academic Press, London.
Graham-Jones, O (1964) *Small animal anaesthesia.* Pergamon Press, Oxford.
Green, C J (1979) *Animal anaesthesia.* Laboratory Animal Handbooks 8. Laboratory Animals, London.

LAC Collected Papers Vol. 6. (1957) *Humane techniques in the laboratory.* Medical Research Council Laboratory Animals Centre.
Lane, D. R. (1980) *Jones's animal nursing* 3rd edn. Pergamon Press, Oxford.
MAFF (1978) *Manual of veterinary investigation laboratory techniques*. 2nd edn. Part 1 Basic techniques, Part 2 Bacteriology, Part 3 Biochemistry, Part 4 Haematology, Part 5 Histology, Part 6 Mycology, Part 7 Parasitology, Part 8 Virology. MAFF Publications, ADAS, Surrey.
Mahoney, R (1966) *Laboratory techniques in zoology.* Butterworth, London.
UFAW (1978) *Humane killing of animals*, 3rd edn. UFAW, London.
UFAW (1976) *The humane destruction of unwanted animals*. UFAW, London.
UFAW (1969) *The humane control of animals living in the wild.* UFAW, London.

Disease

Benirschke, K, Garner, F M & Jones, T C eds. (1978) *Pathology of laboratory animals* Vols. 1 & 2. Springer-Verlag, New York.
Busvine, J R (1975) *Arthropod vectors of disease*. Arnold, London.
Boycott, J A (1971) *Natural history of infectious disease*. Arnold, London.
Brander, G C & Ellis, P R (1977) *The control of disease.* Ballière & Tindall, London.
Chandler, A C (1955) *Introduction to parasitology.* Wiley, New York.
Cotchin, E & Roe, F J C (1967) *Pathology of laboratory rats and mice*. Blackwell, Oxford.
Flynn, R J (1973) *Parasites of laboratory animals.* Ames, Iowa State University Press.
Hime, J M & O'Donoghue, P N eds. (1979) *Handbook of diseases of laboratory animals*. Heinemann, London.
Lapage, G (1968) *Veterinary parasitology*. Oliver & Boyd, Edinburgh.
Owen, D (1972) *Common parasites of laboratory rodents and lagomorphs*. Medical Research Council Laboratory Animals Centre Handbook 1 HMSO.
Roberts, D C (1969) *Index of transplanted tumours of rats and mice.* Laboratory Animal Handbooks 3. Laboratory Animals, London.
Smyth, J D (1976) *Introduction to animal parasitology,* 2nd edn. Hodder & Stoughton, London.
Wilson, R A (1967) *An introduction to parasitology*. Arnold, London.

Health & Safety

Andrewes, C H & Walton, J R (1977) *Viral and bacterial zoonoses*. Ballière, Tindall, London.
Fiennes, R (1967) *Zoonoses of primates*. Weidenfeld & Nicolson, London.
LAC Collected Papers Vol. 10 (1961) *Hazards of the animal house.* Medical Research Council Laboratory Animals Centre.
Perkins, F T & O'Donoghue, P N eds. (1969) *Hazards of handling simians.* Laboratory Animal Handbooks 4. Laboratory Animals, London.
Seamer, J H & Wood, M eds. (1981) *Safety in the animal house*, 2nd edn. Laboratory Animal Handbooks 5. Laboratory Animals, London.

Biology

Abercrombie, M ed. (1977) *A dictionary of biology*. Penguin, London.

Allen, B L (1979) *Basic anatomy. A laboratory manual*. The human skeleton/the cat, 2nd edn. Freeman, San Francisco.

Altman P L & Dittmer D S eds. 1974 *Biology data handbook* 2nd ed. vol. 3. Federation of American Societies for Experimental Biology

Baker, H J, Lindsay, J P & Weisbroth, S H eds. (1979) *The laboratory rat*, Vol. I, Biology and dissection; Vol II, Research applications. Academic Press, New York.

Barrass, R (1979) *Modern biology made simple*. Allen, London.

Clegg, P C & Clegg, A G (1975) *Biology of the mammal*, 4th edn. Heinemann, London.

Cooper, G & Schiller, A L (1975) *Anatomy of the guinea -pig*. Harvard University Press, Cambridge, USA.

Dictionary of biology (1980) Key Facts. Seymore Press Ltd., London.

Falconer, D S (1975) *Introduction to quantitative genetics*. Longman, London.

Freeman, W H & Bracegirdle, B (1963) *An atlas of embryology*. Heinemann, London.

Freeman, W H & Bracegirdle, B (1966) *An atlas of histology*. Heinemann, London.

Green, E L ed. (1966) *Biology of the laboratory mouse*. McGraw-Hill, London.

Mackean, D G (1978) *Introduction to biology*. Murray, London.

Roberts, M B V (1976) *Biology: A functional approach*, 2nd edn. Nelson, Sunbury on Thames.

Robinson, R (1971) *Genetics for cat breeders*. Pergamon Press, Oxford.

Sirockin, G & Cullimore, S (1969) *Practical microbiology*. McGraw Hill, Maidenhead.

Wagner, J E & Manning, P J (1976) *The biology of the guinea -pig*. Academic Press, London.

Weisbroth, S H, Flatt, R E & Kraws, A L (1974) *The biology of the laboratory rabbit*. Academic Press, London.

Wells, T A G (1964) *The rat, a practical guide*. Heinemann, London.

General

Baker, F J & Silverton, R E eds. (1976) *An introduction to medical laboratory technology*. Butterworths, London.

Batty, J, (1979) *Domesticated ducks and geese*. Saiga Publishing Co. Ltd. Surrey

Farris, E J & Griffiths, J Q (1962) *The rat in laboratory investigation*. Hafner, New York.

Festing, M F W, (1979) *Inbred strains in biomedical research*. Macmillan, London.

Festing, M F W, (1980) *International index of laboratory animals*, 4th edn. Medical Research Council Laboratory Animals Centre.

Hafez, E S E ed. (1969) *The behaviour of domestic animals*. Ballière, Tindall & Cassell, London.

Hobbs, K R & Beby, J (1976) *Laboratory non -human primates for biomedical research in the United Kingdom*. Medical Research Council Laboratory Animals Centre.

Hulme, S (1979) *Book of the pig.* Saiga Publishing Co. Ltd., Surrey.
Inglis, J. K (1980) *Introduction to laboratory animal science and technology.* Pergamon Press, Oxford.
LAC (1974) *Survey of the numbers and types of laboratory animals used in the United Kingdom in 1972.* LAC Manual Series No. 3. Medical Research Council Laboratory Animals Centre.
LAC (1974) *The accreditation & recognition schemes for suppliers of laboratory animals.* 2nd edn. Medical Research Council Laboratory Animals Centre. Manual Series No. 1.
LAC Collected Papers Vol. 7 (1958) *The organisation and administration of an animal division.* Medical Research Council Laboratory Animals Centre.
LAC Collected Papers Vol. 8 (1959) *Quality in laboratory animals.* Medical Research Council Laboratory Animals Centre.
LAC Collected Papers Vol. 9 (1960) *The provision of animals for cancer research.* Medical Research Council Laboratory Animals Centre.
LAC Collected Papers Vol. 12 (1963) *The choice of experimental animal.* Medical Research Council Laboratory Animals Centre.
Lane-Petter, W. (1961) *Provision of laboratory animals for research – a practical guide.* Elsevier, London.
Lane-Petter, W & Pearson, A E G (1971) *The laboratory animal – principles and practice.* Academic Press, London.
McKenzie, D (1980) *Goat Husbandry*, 4th edn. Faber & Faber, London.
Morse, H C (1978) *Origins of inbred mice.* Academic Press, London.
Mount, L E & Ingram D L (1971) *The pig as a laboratory animal.* Academic Press, London.
Parker, R E (1973) *Introductory statistics for biology.* Arnold, London.
Parrot, R F & Festing, M F W (1974) *Standardised laboratory animals.* LAC Manual Series No. 2 Medical Research Council Laboratory Animals Centre.
Sandford, J C (1979) *The domestic rabbit*, 3rd edn. Crosby Lockwood Staples, London.
Sainsbury, D (1979) *Poultry health and management.* Granada, St Albans.
Short, D J & Woodnott, D P, eds. (1969) *The IAT manual of laboratory animal practice and techniques*, 2nd edn. Crosby, Lockwood, Staples – Granada, St Albans.
Smyth, D H (1978) *Alternatives to animal experiments.* Scolar Press, London.
UFAW (1969) *The use of animals in toxicological studies.* UFAW, London.
UFAW (1971) *The UFAW handbook on the care and management of farm animals.* Churchill Livingstone, Edinburgh.
UFAW (1972) *The rational use of living systems in biomedical research.* UFAW, London.
UFAW (1973) *The welfare and management of wild animals in captivity.* UFAW, London
UFAW (1976) *The UFAW handbook on the care and management of laboratory animals*, 5th edn. Churchill Livingstone, Edinburgh.
UFAW (1977) *The welfare of laboratory animals.* UFAW, London.

Appendix 4
Journals, newsletters and useful addresses

Journals and Newsletters

Journal of the Institute of Animal Technicians.
  Mrs J Robertson, Animal Unit, Western General Hospital, Edinburgh.
Laboratory Animals.
  Laboratory Animals Ltd., 33 Furrowfelde, Kingswood, Basildon, Essex, SS15 5HA
Laboratory Animal Science (formerly Laboratory Animal Care).
  American Association of Laboratory Animal Science, 210 North Hammes Avenue, Suite 205, Joiliet, IL 60435.
Guinea-pig News Letter Ed. Dr M F W Festing
  MRC Laboratory Animals Centre, Carshalton, Surrey
Mouse News Letter Ed. Dr A G Searle
  MRC Laboratory Animals Centre, Carshalton, Surrey
Rat News Letter Ed. Dr M F W Festing
  MRC Laboratory Animals Centre, Carshalton, Surrey

Useful Addresses

American Association of Laboratory Animal Science
  College of Veterinary Medicine, Iowa State University, Ames 1A 5011
Animal Health Trust
  Landwades Hall, Kennet, Newmarket, Suffolk CB8 7PN
Her Majesty's Stationery Office (HMSO)
  49 High Holborn, London, WC1
Home Office Inspectorate
  H3 Division, 50 Queen Anne's Gate, London, SW1H 9AT
Institute of Animal Technicians
  5 South Parade, Summertown, Oxford, OX2 7JL
Institute of Biology
  41 Queen's Gate, London, SW7 5HU
Laboratory Animals Science Association
  c/o M W Smith, Director, Cambridge University Central Animal Services, New Addenbrookes Hospital, Hills Road, Cambridge, CB2 2QL
Medical Research Council Laboratory Animals Centre (LAC)
  Woodmansterne Road, Carshalton, Surrey, SM5 4EF
Research Defence Society (RDS)
  Lettsom House, 11 Chandos Street, Cavendish Square, London, W1M 9DE
Universities Federation for Animal Welfare (UFAW)
  8 Hamilton Close, South Mimms, Potters Bar, Herts, EN6 3QD

Index

Printed in Dunstable, United Kingdom

91900929R00137